Optimizing Research Project Management in Education

Elomari Driss
https://orcid.org/0000-0001-5846-8501
Sidi Mohamed Ben Abdellah University, Morocco

Rahma Barbara
Sidi Mohamed Ben Abdellah University, Morocco

Elassaad Elharbaoui
https://orcid.org/0000-0002-1946-3561
Higher Institute of Childhood Professionals, University of Carthage, Tunisia

Jana Birova
Charles University, Czech Republic

Mustapha Khiati
Hassan II University, Morocco

IGI Global
Scientific Publishing
Publishing Tomorrow's Research Today

Vice President of Editorial — Melissa Wagner
Director of Acquisitions — Mikaela Felty
Director of Book Development — Jocelynn Hessler
Production Manager — Mike Brehm
Cover Design — Jose Rosado

Published in the United States of America by
IGI Global Scientific Publishing
701 East Chocolate Avenue
Hershey, PA, 17033, USA
Tel: 717-533-8845 | Fax: 717-533-7115
Website: https://www.igi-global.com E-mail: cust@igi-global.com

Library of Congress Cataloging-in-Publication Data

Names: Elomari, Driss, 1981- editor http://id.loc.gov/vocabulary/relators/edt http://id.loc.gov/authorities/names/n2025011491 http://id.loc.gov/rwo/agents/n2025011491
Title: Optimizing research project management in education / [edited by] Driss Elomari, Barbara Rahma, Elharaoui Elassaad, Jana Birova, Mustapha Khiati.
Description: Hershey, PA : IGI Global Scientific Publishing, [2026] | Includes bibliographical references and index.
Identifiers: LCCN 2025018912 (print) | LCCN 2025018913 (ebook) | ISBN 9798337329772 ebook | ISBN 9798337329758 hardcover | ISBN 9798337329765 paperback
Subjects: LCSH: Education--Research http://id.loc.gov/authorities/subjects/sh85041016 | Project management http://id.loc.gov/authorities/subjects/sh85065919
Classification: LCC LB1028 (ebook) | LCC LB1028 .O68 2026 (print) | DDC 370.72 23/eng/20250--dc26
LC record available at https://lccn.loc.gov/2025018912
British Cataloguing in Publication Data
A Cataloguing in Publication record for this book is available from the British Library.

Editorial Advisory Board

Table of Contents

Detailed Table of Contents

Strategic planning is the quiet engine behind every successful educational research project. This chapter explores how deliberate, thoughtful planning transforms a conceptual idea into a fully realized research endeavor with meaningful outcomes. By bridging the gap between intention and execution, the chapter examines how clarity of purpose, methodological alignment, resource foresight, and adaptive responsiveness shape the trajectory of research. It also reflects on the often-overlooked emotional and intellectual labor involved in navigating uncertainty, making course corrections, and defining what success truly means in an educational context. Ultimately, the discussion invites researchers to approach planning not as a rigid protocol but as a living process—one that honors complexity, fosters innovation, and ensures that research efforts contribute sustainably to the evolving landscape of education.

In the Moroccan educational system, the French-language proficiency of middle school students remains poor, particularly in vocabulary and communication. The memorization-based traditional methods do not help develop actual language practice. The present article describes an innovative pedagogical project aimed at integrating theater into teaching French to improve the language skills of students. Theater, being an interactive game like activity, provides vocabulary development and communication fluency. Participating in theater activities, such as role-playing and improvisations, allows students to learn in a dynamic and contextualized way. Our project had positive outcomes, especially in improving students; oral and written skills. Students became more interested in learning French, which also boosted their motivation and confidence. This teaching method proves to be a valuable and flexible option for addressing the difficulties of teaching French in Morocco.

This chapter investigates tokenization as an innovative mechanism for managing educational research projects. Rooted in blockchain technology, it provides for transparent tracking, recognition, and motiving of individual contributions via fungible and non-fungible tokens. The proposed model integrates smart contracts (to automate much of the work), collaborative governance, and real-time monitoring. Case studies like BitDegree and ResearchHub illustrate how tokenization can enhance fairness, engagement, and accountability, and we address a number of ethical, technical, and pedagogical issues, too. These are important because they're about aligning educational research with its core values

This study presents an exploration of operational research, transport, and logistics in the healthcare sector, highlighting their critical role in optimizing resource allocation, improving service delivery, and enhancing patient outcomes. Efficient healthcare logistics ensure the timely distribution of medical supplies, pharmaceuticals, and life-saving equipment, particularly in emergency response and remote areas. Operational research techniques, including mathematical modeling, simulation, and data analytics, help streamline supply chains, reduce costs, and minimize wastage. Additionally, advancements in technology, such as artificial intelligence (AI), real-time tracking, and predictive analytics, are transforming healthcare logistics by enhancing efficiency and responsiveness. Challenges such as infrastructure limitations, regulatory barriers, and disruptions due to pandemics or natural disasters continue to hinder seamless healthcare logistics. This study examines innovative strategies and best practices in operational research and transport management for sustainability.

We present a case study in the management of an educational research project through the design of an Arabic text simplification system. The push towards AI implementations within education has produced a parallel demand for personalized, accessible content to accommodate language learners. This research project builds towards addressing these gaps, by designing and developing a lexical simplification framework that combines linguistic resources with the latest advances in deep learning models. It takes the reader through the project lifecycle while also noting the overall decisions, methodological innovations and management strategies made throughout the project lifecycle. The system produces identifications of complex words and simplified synonym options based on an innovative combination of the Arabic WordNet, lexicons manually composed, word embeddings and the pre-trained AraBERT model. The outcomes are assessed using a combination of text readability classification and semantic similarity, to produce quantifiable improvements in readability while sufficiently retaining meaning.

This article covers a problematic concerning most of Moroccan learners of French as a foreign language. It covers one of the four main skills namely writing. Based on a test covering the writing of an intimate letter, this study focuses mainly on the linguistic approach and precisely the enunciative approach. The goal is to see the usage rate of enunciative aspects in students' writings especially since the enunciative system require the presence of several elements. This work aims to show if learners can exploit language course in their writing and if they correctly

Preface

The landscape of education is undergoing a profound transformation, driven by rapid technological advancements, evolving pedagogical methodologies, and an increasing demand for data-driven decision-making. In this dynamic environment, the effective management of research projects has emerged as a critical factor in advancing educational practices and outcomes. Optimizing research project management in education is not merely an administrative necessity—it is a strategic imperative capable of enhancing the quality, relevance, and impact of educational research.

Research in education serves as the cornerstone for developing innovative teaching methods, improving learning outcomes, and addressing the diverse needs of students. Yet the complexity and multidimensional nature of educational research projects present unique challenges. Effective management must encompass project planning, resource allocation, stakeholder engagement, and the dissemination of research findings to ensure projects are completed on time, within budget, and to the highest standards of quality.

This book, *Optimizing Research Project Management in Education*, provides a comprehensive guide to navigating the intricacies of managing educational research projects. It integrates theoretical foundations with practical applications of project management principles tailored to the educational context. By drawing on insights from project management, educational research, and data analytics, this volume offers a holistic approach to enhancing research project outcomes.

The chapters explore key dimensions such as strategic planning, collaboration and communication, the use of technology in project management, and the evaluation and dissemination of research outcomes. Each chapter equips educators, researchers, and administrators with practical knowledge and tools to manage research projects effectively, thereby contributing to the advancement of educational practices and policies.

In an era where data and technology increasingly shape the future of education, the ability to manage research projects efficiently is more crucial than ever. This book aims to empower educators and researchers to harness project management principles to drive meaningful change and innovation. By adopting the strategies outlined here, readers can enhance their capacity to conduct impactful research, ultimately benefiting students, institutions, and the broader educational community.

OBJECTIVES OF THE BOOK

The primary objective of this volume is to provide comprehensive and practical guidance for managers, researchers, and administrators seeking to optimize research projects within the educational context. It seeks to:

- **Enhance Understanding of Project Management Principles:** Equip readers with a foundation in project management tailored to education, including strategic planning, resource allocation, risk management, and stakeholder engagement.
- **Provide Practical Tools and Strategies:** Offer actionable methodologies, best practices, and guidance on leveraging technology and data analytics to optimize project outcomes.
- **Promote Data-Driven Decision Making:** Highlight the importance of collecting, analyzing, and interpreting data effectively to inform research decisions and increase project impact.
- **Foster Collaboration and Communication:** Emphasize teamwork, stakeholder engagement, and knowledge sharing to ensure research projects are aligned with the needs of the educational community.
- **Advance Current Research:** Contribute new insights, methodologies, and case studies that bridge theory and practice in the management of educational research projects.
- **Support Professional Development:** Serve as a resource for educators, researchers, and administrators to enhance their project management skills and drive meaningful change in education.

INTENDED AUDIENCE

This book is designed to serve a diverse audience within the educational and research communities, including:

- **Educational Researchers:** Scholars and practitioners engaged in research will gain tools and strategies for effective project management, from inception to dissemination.

- **Teachers and Educators:** Practitioners at all levels can apply insights to improve teaching practices, integrate research findings, and support curriculum development.
- **Administrators and Policymakers:** Leaders responsible for research oversight and policy implementation can better allocate resources, manage projects efficiently, and align outcomes with institutional goals.
- **Graduate Students:** Students in education, educational research, or related fields will acquire foundational project management knowledge applicable to research, teaching, and administrative roles.
- **Professional Developers and Trainers:** Those designing and delivering training programs can use this book to enhance participants' project management competencies.
- **Data Analysts and Technologists:** Professionals in educational technology and analytics will benefit from insights on leveraging technology and data to optimize research management.

In summary, *Optimizing Research Project Management in Education* offers a comprehensive, practical, and research-informed guide to managing educational research projects. It empowers a wide range of professionals to enhance the quality, efficiency, and impact of their research initiatives, ultimately contributing to the advancement of educational practices and policies. We invite you to explore this volume and discover how optimizing research project management can lead to more effective, innovative, and impactful educational research.

ORGANIZATION OF THE BOOK

Chapter 1: Strategic Planning in Educational Research: From Concept to Completion

This chapter emphasizes the centrality of strategic planning in educational research. It portrays planning not as a rigid checklist but as a dynamic, living process that bridges conceptual ideas and actionable research outcomes. The discussion highlights how clarity of purpose, methodological alignment, resource foresight, and adaptive responsiveness shape project trajectories. Beyond technical considerations, the chapter also acknowledges the emotional and intellectual labor inherent in navigating uncertainties, making course corrections, and defining success. Researchers are invited to approach planning as a tool for fostering innovation and sustaining meaningful contributions to educational practice.

Chapter 2: Theatrical Innovations and Research Laboratory in French Class

Focusing on language acquisition in Moroccan middle schools, this chapter explores the integration of theater into French language instruction. Moving away from traditional memorization-based methods, the chapter demonstrates how interactive, role-playing activities enhance vocabulary development, communication fluency, and overall student engagement. The outcomes reveal improved oral and written skills, increased motivation, and heightened confidence among students, offering a flexible, culturally relevant pedagogical strategy for language education.

Chapter 3: Enhancing Resource Management Efficiency with Technology Integration on Educational Research Projects

This chapter examines how technology and strategic resource allocation improve the management of educational research projects in Moroccan higher education institutions. Through a case study approach, it explores challenges related to budgetary limits and human resources while highlighting benefits derived from project management software and data analytics tools. Emphasizing strategic planning, stakeholder mobilization, and risk management, the chapter demonstrates how technological integration strengthens teamwork, communication, and overall research efficiency, offering transferable insights for educational institutions worldwide.

Chapter 4: Towards Excellence in E-Learning: Implementing SEQCL for Course Quality

Addressing the transformation of course delivery in the digital era, this chapter introduces the SEQCL platform (Quality Evaluation System of Online Courses) as a tool for enhancing e-learning quality. It provides an overview of quality criteria for online course design and presents a practical case study involving 22 instructors. The findings illustrate how structured guidance, combined with training in digital tools like Moodle, supports the development of effective, learner-centered online courses that meet established pedagogical standards.

Chapter 5: Tokenizing Educational Research: Toward Incentive-Based and Decentralized Project Governance

Exploring blockchain-based innovations, this chapter investigates tokenization as a mechanism for incentivizing and managing contributions in educational research. By integrating smart contracts, collaborative governance, and real-time monitoring, the chapter demonstrates how tokenization can promote transparency, engagement, and accountability. Case studies such as BitDegree and ResearchHub illustrate both the potential and challenges of aligning research governance with ethical and pedagogical principles, offering a forward-looking perspective on decentralized project management.

Chapter 6: Operational Research, Transport, and Logistics in the Healthcare Sector

This chapter highlights the pivotal role of operational research, transport, and logistics in optimizing healthcare delivery. It examines strategies to streamline supply chains, allocate resources efficiently, and enhance service quality through techniques like mathematical modeling, simulation, and data analytics. The chapter also considers technological advancements, including AI and predictive analytics, and addresses persistent challenges such as infrastructure limitations, regulatory barriers, and crisis-driven disruptions, offering best practices for sustainable healthcare logistics.

Chapter 7: Managing Educational Natural Language Processing Projects with Hybrid Artificial Intelligence

Focusing on Arabic text simplification, this chapter presents a case study on managing AI-driven educational research projects. It details the development of a hybrid system combining linguistic resources with deep learning models, illustrating how complex words can be identified and simplified without losing semantic meaning. The chapter emphasizes project lifecycle management, methodological innovation, and AI integration, demonstrating quantifiable improvements in text readability and offering a model for future AI-enhanced educational research projects.

Chapter 8: Managing Educational Projects through the Enunciative Exploration of High School Students' Writings

This chapter examines Moroccan students' mastery of writing skills in French as a foreign language, with a focus on enunciative elements. Through analysis of intimate letters, it identifies common challenges in students' application of linguistic and enunciation markers. The findings reveal gaps in comprehension and usage, highlighting the importance of targeted pedagogical interventions to enhance written communication skills and ensure the effective transfer of classroom learning into practical writing competence.

CONCLUSION

As we conclude this volume, we are struck by the richness and diversity of insights presented across the chapters, each offering a unique lens on the challenges and opportunities inherent in managing educational research projects. From strategic planning and resource allocation to technological integration, blockchain-based governance, AI-driven innovations, and collaborative learning, the chapters collectively highlight that effective research project management is both a science and an art—requiring methodological rigor, adaptive strategies, and a deep understanding of human, institutional, and technological factors.

A central theme emerging from this volume is that the success of educational research projects depends not only on technical competence but also on thoughtful leadership, ethical considerations, and a collaborative spirit. Whether through theater-based pedagogical interventions, e-learning quality frameworks, or systematic reviews of project management in entrepreneurship education, each contribution underscores the importance of aligning research objectives with practical, student-centered outcomes. The chapters demonstrate that innovation in education is most impactful when it is guided by clear planning, robust resource management, and inclusive, participatory approaches that respect the diversity of learners, educators, and stakeholders.

This book also emphasizes the transformative potential of technology in educational research. From AI-driven text simplification systems to integrated digital collaboration frameworks and tokenized research governance, the chapters illustrate how digital tools can enhance efficiency, transparency, and engagement. Yet, these innovations are most effective when integrated thoughtfully, with careful attention to ethics, accessibility, and the human dimension of education.

Ultimately, *Optimizing Research Project Management in Education* seeks to equip educators, researchers, administrators, and policymakers with the knowledge, strategies, and practical tools necessary to navigate the complex landscape of educational research. We hope this volume inspires readers to adopt a holistic and reflective approach to project management—one that balances rigor with creativity, structure with flexibility, and technology with human insight. By doing so, we believe educational research can achieve its fullest potential, generating meaningful, sustainable impacts on teaching, learning, and the broader educational community.

Chapter 1
Strategic Planning in Educational Research: From Concept to Completion

Mustafa Kayyali
https://orcid.org/0000-0003-3300-262X
Maaref University of Applied Sciences, Syria

ABSTRACT

Strategic planning is the quiet engine behind every successful educational research project. This chapter explores how deliberate, thoughtful planning transforms a conceptual idea into a fully realized research endeavor with meaningful outcomes. By bridging the gap between intention and execution, the chapter examines how clarity of purpose, methodological alignment, resource foresight, and adaptive responsiveness shape the trajectory of research. It also reflects on the often-overlooked emotional and intellectual labor involved in navigating uncertainty, making course corrections, and defining what success truly means in an educational context. Ultimately, the discussion invites researchers to approach planning not as a rigid protocol but as a living process—one that honors complexity, fosters innovation, and ensures that research efforts contribute sustainably to the evolving landscape of education.

INTRODUCTION

Educational research, at its core, is a quest for understanding—an ever-evolving effort to illuminate the practices, policies, and principles that shape learning in diverse contexts. But like any meaningful journey, the process of research requires

DOI: 10.4018/979-8-3373-2975-8.ch001

more than curiosity or intellectual passion. It demands structure. It demands foresight. It demands planning.

Yet, strategic planning in educational research is not merely a bureaucratic formality or a checklist exercise to satisfy funding bodies and institutional review boards (Caeymaex et al, 2023). It is a vital act of intellectual architecture—an intentional mapping of purpose, possibility, and process. Without a solid strategic plan, even the most promising research questions can drift into ambiguity, and projects that begin with conviction may falter under the weight of competing demands, shifting priorities, and inevitable constraints.

In today's educational environment—marked by rapid technological advancement, political contestation, and social upheaval—the stakes for rigorous and impactful research have never been higher (Patton, 2022). Researchers are not only called upon to produce knowledge but to ensure that their work is ethically sound, methodologically robust, contextually relevant, and ultimately transformative. Strategic planning serves as the scaffolding for that responsibility. It enables scholars to design research projects that are coherent yet flexible, ambitious yet grounded in reality. But what does "strategic" truly mean in this context? It is not simply about efficiency or project completion. Rather, strategic planning involves asking hard questions at the outset: Why this research? Why now? Who will it serve? What will it leave behind? It invites researchers to think holistically—not only about methods and timelines, but also about values, relationships, impact, and legacy. It encourages researchers to anticipate challenges without succumbing to rigidity, and to embrace complexity without becoming paralyzed by it.

This chapter unfolds with a single guiding conviction: that the planning phase of educational research is not ancillary to the work—it *is* the work. The decisions made at the conceptual stage ripple outward, shaping everything from data collection to interpretation, dissemination, and use. Whether working on a classroom-based action research project or a longitudinal policy study, researchers must see planning not as a static prelude, but as a dynamic, ongoing process that evolves with insight, circumstance, and reflection. We begin by revisiting the often-neglected starting point: the act of naming a problem and framing a meaningful question. We then explore how methodological decisions can serve as extensions of strategic thinking, not isolated technical choices. Further sections examine resource management, adaptive responsiveness, and evaluation strategies—all with the aim of helping researchers build a blueprint that is as thoughtful as the inquiry itself.

Ultimately, this chapter offers more than procedural advice. It advocates for a mindset shift: from viewing planning as a chore to embracing it as an act of intellectual and ethical stewardship. For in educational research, as in life, the most meaningful outcomes rarely emerge by accident—they are shaped by careful, courageous, and creative planning.

LAYING THE FOUNDATION: DEFINING PURPOSE AND FRAMING RESEARCH QUESTIONS

Every research journey begins in silence. Before the literature review. Before the theoretical scaffolding. Before any ethics applications or methodologies are considered (Schweiger & Tomiak, 2022). There is that brief, almost unspoken moment—something between a hesitation and a whisper—when the researcher sits with a feeling, a hunch, a discomfort, a curiosity. It's not quite a problem yet, not yet a question. Just a pulse, a disturbance in the intellectual air. That's where everything starts. And it's here, in this formless beginning, that the most consequential decisions are made. The kind of decisions that don't always announce themselves as pivotal but, over time, dictate everything that follows. Not just the direction of the research, but the kind of energy it will carry. The kind of questions it will allow. The kind of impact it might eventually have. Strategic planning doesn't begin with scheduling or budgeting. It begins with a reckoning—with purpose (Dodson, 2023). But purpose is not a neat thing. It doesn't arrive prepackaged in a mission statement or a polished paragraph. It has to be dug out. Excavated, almost. Like something buried beneath years of academic habit, institutional pressures, and the need to sound clever. In educational research especially, where issues are tangled with social justice, human growth, power, and pain, the purpose of a study is never just about filling a gap in the literature. It's about deciding where to stand. Who to listen to. What to fight for.

That's the thing about framing purpose: it's political, even when it pretends not to be. So, how do we begin? How does one move from that hazy feeling of "something's not quite right" to a research question that can bear the weight of an entire project? We begin by sitting with the discomfort. Letting it speak. Sometimes, the purpose of a study comes from a deeply personal place—a memory of exclusion, a moment of awe, a story overheard in the corridor of a crumbling school. Other times, it's sparked by policy shifts, funding pressures, or institutional agendas. But whatever its origin, purpose becomes the compass. Not the map—those can change—but the quiet pull of north. It tells us not only what we're trying to understand, but why it matters. And that "why" is never neutral. It's soaked in values.

A common misstep—especially among novice researchers, but by no means exclusive to them—is to conflate purpose with topic (Oshchepkova, 2023). "I'm doing research on online learning." That's not purpose. That's terrain. Purpose is what you hope to change, expose, or understand within that terrain. "I want to explore how marginalized students experience disconnection in online spaces." Now we're talking. Now there's a heartbeat. But even that's just the start. Because purpose demands specificity. It asks the researcher to take sides, to locate themselves not just in relation to the literature, but to the people, structures, and histories that surround the issue. You can't frame a meaningful research question until you've confronted your own position in the story—your assumptions, your blind spots, your biases. This is where strategic planning gets personal.

Ask yourself: What's driving me toward this inquiry? What am I hoping will happen as a result of it? Who stands to gain? Who might be harmed? Who hasn't been heard yet? These aren't just ethical questions—they're strategic ones. Because every research decision flows downstream from how we define our purpose. Then comes the art of translation: taking that raw purpose and shaping it into a research question. This is harder than it looks. A good research question doesn't just summarize your interests—it challenges them. It refuses to let you stay comfortable (Alvesson & Sandberg, 2024). It opens up complexity rather than shutting it down. And it does all of that while still being researchable. Too broad, and you drown in generalities. Too narrow, and you lose sight of the bigger picture. Too abstract, and no data can reach it. Too concrete, and it loses meaning. Framing the right question is less like filling out a form and more like tuning an instrument. You keep adjusting—slightly, subtly—until the tone is right. But here's the paradox: the more you refine the question, the more it reveals your values. That's not a flaw. That's the point.

In educational research, questions are never just technical tools—they are acts of commitment (Maki, 2023). When you ask, "How do teachers navigate conflicting pedagogical demands in underfunded schools?" you are already saying something about what you believe matters. When you ask, "What supports resilience in refugee students navigating disrupted schooling?" you are not just seeking information—you are shaping attention, inviting empathy, demanding nuance. There is no such thing as a neutral research question. And strategic planning means owning that.

Let's not pretend this part of the process isn't messy. It's supposed to be. Framing purpose and questions requires a willingness to move through uncertainty, to question your first instincts, to resist the seductive clarity of easy answers (Lehman, 2024). Sometimes you'll write a draft question and sit with it for days, only to throw it out because it feels too tidy, too performative, too far from the messiness that first sparked your interest. That's not a waste of time—it's a necessary sifting. And while templates and taxonomies may help—those checklists that distinguish between "descriptive," "explanatory," "evaluative," and "critical" questions—they can also flatten your inquiry if you lean on them too early. Real insight comes not from categorizing your question but from interrogating it. Who does this question exclude? What assumptions does it carry? Is it asking the easy version of a hard thing?

At some point, you'll realize you're not just crafting a research question—you're crafting a doorway. A portal into a world of data, decisions, dilemmas. If you frame it well, it will lead you not only to answers but to better, deeper questions. Strategic planning, at this stage, means resisting the urge to rush (Kayyali, 2025 A). Take the time to write multiple versions of your question. Test them in conversation. Imagine what kinds of data they invite—and what kinds they silence. Don't just ask if a question is "feasible." Ask if it's worthy. Because the truth is, a lot of educational research gets done that is feasible but forgettable. Safe but shallow. Methodologically tight but morally timid. Strategic planning—done with care—can help us avoid that (Finn & Wylie, 2021). It can help us find the questions that unsettle us just enough to keep us honest, and that anchor us just enough to keep us steady.

It's also worth saying: you don't have to do this alone. Purpose and questions grow stronger in dialogue. With mentors, peers, communities, sometimes even with those we seek to research. Strategic planning thrives on feedback—on resistance, on pushback, on someone saying, "But have you thought about it this way?" Planning is not an individual act of genius. It's a relational practice. And it evolves. The question you start with may not be the one you end with. That's not failure. That's learning. But beginning with intention—with a deep, well-considered purpose—gives you a center of gravity. Something to return to when the data is murky, the deadlines loom, and the confidence falters.

Let's not underestimate how powerful that center of gravity can be. In the chaotic ecosystem of academic research—with its shifting theories, complex ethics, and ever-demanding deliverables—your clarity of purpose is what holds you (Sarno et al, 2025). It's the thing that makes the whole project more than just an intellectual exercise. It makes it a contribution. A good question leads somewhere. But a great question? It lingers. It reverberates beyond the study. It changes how people see, think, feel. It outlasts your funding cycle. It creates space—for action, for policy, for healing, for change.

So, let's return, one last time, to that quiet moment at the beginning. When the world hasn't yet demanded outputs or citations. When there's just a whisper of wondering. Honor that moment. Sit with it. Let it deepen. That's where your research lives. Because before anything is measured, coded, or analyzed—before the tools and tables and timelines—there is the question. And the question, if you've chosen it well, will ask something of you too. Not just what you want to know. But who you're willing to become in the process of finding out.

DESIGNING FOR IMPACT: STRATEGIC ALIGNMENT OF METHODOLOGY AND GOALS

There's something oddly comforting about method sections. They feel clean, right? Predictable. A zone of control in the otherwise stormy process of research. Lists of instruments, procedures, sample sizes—like furniture in a well-arranged room. A place where certainty pretends to live. But the truth is, methodology is anything but neutral. It's not just a technical detail or an obligatory chapter in the dissertation template. It's a decision—layered, political, revealing. A decision that tells the world how you think knowledge works, what you believe counts as evidence, and whose voices deserve to be recorded, interpreted, and amplified (Baines et al, 2023). And if you're not careful, it's also where your purpose goes to die. Because here's the thing: misaligned methodology is the silent killer of meaningful research. You can have the most urgent purpose, the most piercing research question, but if your method doesn't match your intent—if it contradicts the logic of your inquiry or flattens the complexity you're trying to explore—then your findings will feel hollow. Worse, they'll mislead.

Designing for impact means taking the risk of letting purpose shape method—not the other way around. Too often, researchers reverse-engineer their studies from familiar methodological templates. Surveys because they're easy to scale. Interviews because they feel personal. Case studies because they've seen them done before. But starting with a method and trying to stretch your research question to fit it is like tailoring a suit without measuring the body—it might look sharp from afar, but it won't move with you (Gioia, 2021).

This is where strategic planning enters the room—not as bureaucracy, but as discipline. As intentionality. The kind of careful attention that asks: What do I want this research to *do* in the world? Not just academically, but practically, socially, ethically. And then, with equal seriousness: What kind of design will help me do that?

Strategic alignment between methodology and goals doesn't mean forcing harmony (Salman, 2023). It's not about making every part of the study symmetrical or balanced. In fact, the most impactful research often lives in the tension between messiness and method, between aspiration and feasibility. It's about coherence, not perfection. Integrity, not polish. Let's take a step back. What does "impact" really mean in educational research?

We toss the word around as if its meaning is self-evident, but it's slippery. For some, impact is about policy influence—a study that changes how a ministry funds schools. For others, it's about pedagogical insight—a study that transforms a teacher's practice in a real classroom. Sometimes impact is subtle: a shift in discourse, a reframing of assumptions, a new metaphor that helps someone see differently. Sometimes it's personal. Sometimes it's systemic.

The point is, impact doesn't always look like a headline or a citation spike. It often whispers instead of shouting. But designing for impact means planning as if someone else's life, practice, or understanding will change because of what you've found. It's not about ego—it's about responsibility. And that responsibility extends to your method. If your research question is rooted in lived experience, in inequality, in cultural nuance—then choosing a method that quantifies but does not contextualize can be a betrayal. If your goal is to listen deeply, then methods that privilege speed and surface-level responses can drown out the voices you were trying to center. It's not that any one method is "bad." It's that some methods, when misaligned, distort the very thing they claim to illuminate.

Let's take a concrete example. Suppose your purpose is to explore how refugee students navigate feelings of belonging in host-country schools. That's not a simple variable to measure. It's layered with trauma, language, power, identity. You could distribute a Likert-scale questionnaire that asks, "On a scale from 1 to 5, how welcome do you feel in your classroom?" That would produce data—numbers, patterns, correlations. But would it truly capture the nuanced, fluctuating, often contradictory emotional landscapes of those students? Or would it compress their humanity into digestible but meaningless averages?

Now imagine instead a participatory approach—student-led photo narratives, for instance, where they document spaces of safety and alienation. Or a series of narrative interviews conducted in the student's first language, analyzed not just for themes but for silences. The difference isn't just methodological. It's ethical. It's about who is allowed to speak, how, and in what form their insights are taken seriously.

Strategic planning means sitting with those questions early—not after you've committed to a design, but at the point where design and purpose are still holding hands, still able to influence one another (Paddeu et al, 2024). It means resisting the pressure to adopt the "expected" method just because it's familiar or institutionally favored. Sometimes, designing for impact means choosing a harder path (Kayyali, 2025 B). A less publishable one. One that doesn't fit neatly into traditional journals but honors the complexity of the issue.

It also means thinking beyond data collection. Strategic alignment requires thinking about the entire arc of the study—from design to dissemination (Abuhassna et al, 2024). Will your findings be legible to the people they most concern? Will they be actionable? Will they speak truth to power—or just echo existing hierarchies? If your study is meant to challenge injustice but your method reinscribes authority, then something has gone awry.

We also have to talk about scale. Big data doesn't always mean big insight. Small studies, when strategically designed, can ripple outward in powerful ways (Pauwels & Aksehirli, 2025). A single ethnographic study of a classroom might not yield generalizable results, but it can expose cracks in dominant assumptions. It can challenge the "what works" discourse by showing how "what matters" is something altogether different.

There's no formula here. No master list of perfect pairings between method and goal. Strategic alignment is more art than algorithm. But there are questions—sharp, persistent questions—that help keep us honest:

- Who is at the center of this inquiry?
- What counts as knowledge here?

- What is being left out by the design I'm considering?
- Am I designing to confirm something I already believe, or to discover something I don't yet understand?

Those aren't checklist items. They're checkpoints of integrity. And then there's the question of time. Impact doesn't always fit into semester schedules or funding cycles. Sometimes, the design that allows for deep insight takes longer than is convenient. Sometimes, a strategically aligned study asks you to linger, to return, to build relationships that can't be rushed. That's where strategic planning collides with institutional realities. It's tempting to cut corners—to simplify, accelerate, extract. But those shortcuts come at a cost. Because educational research is never neutral. It happens in real places, with real people. And every methodological choice carries consequences.

This is why strategic design must also be transparent. Researchers need to show their work—not just what they did, but why they did it that way. Explain the trade-offs. Acknowledge the constraints. Own the choices. Strategic alignment is not about pretending to have found the perfect method (Hubinger et al, 2024). It's about demonstrating that you cared enough to choose deliberately. In fact, some of the most compelling research I've read came from projects where the authors admitted where their method didn't fully fit, where they had to adapt, where unexpected realities forced revisions. That vulnerability, that openness—it's part of strategic planning too. Because it tells the reader: this was real. This wasn't just methodological theater. This was work that lived in the world. And speaking of the world—don't forget to return to it. As you design, ask not only whether your method will yield results, but whether it will yield relevance. Will it speak to practitioners, to communities, to policymakers, to students? Will it resonate beyond the academic bubble?

Strategic alignment, at its best, is an act of hospitality. It invites others in. It says: this study wasn't just built for publication. It was built to matter (Evans, 2024). And sometimes, strangely enough, the most strategic thing you can do is to choose a method that doesn't guarantee clean answers. A method that makes space for contradiction, for complexity, for voices that don't resolve into themes. Because sometimes, what a field needs isn't more data. It's more courage. Courage to design research that doesn't just mirror the dominant logic of academia but pushes against it. Courage to admit that the method you love might not serve the goal you've set. Courage to ask what impact really means—and then design accordingly. That's what strategic alignment looks like. Not efficiency. Not elegance. But integrity. So, as you design your study, don't just ask: "What method should I use?" Ask: "What kind of world am I helping to build through this work?" And then, build backwards from that.

MAPPING THE JOURNEY: TIMELINES, MILESTONES, AND RESOURCE ALLOCATION

There's a peculiar magic to beginnings. When a research project is still an idea—clean, expansive, unconstrained by time or cost—it can feel limitless. Like standing on a cliffside and staring out across the ocean, full of promise. But then comes the real work: building the boat, charting the route, packing the supplies. You don't just need a dream. You need a plan. And here is where many researchers flinch. Not because they don't care, but because they've been taught to value ideas over infrastructure. The belief lingers that a "real" academic wrestles with theory and epistemology, not timelines or resource spreadsheets. But this is a dangerous illusion. Ideas without structure float. They don't land. Mapping the journey of a research project is not about micromanaging the imagination. It's about creating a rhythm that helps the work breathe without collapsing. It's about anticipating fatigue, celebrating momentum, and knowing when to pause. The best plans don't just schedule tasks. They tell a story. A story of effort, of growth, of unfolding intention.

So, let's begin with the most fragile, most underestimated component: time. Time in research is never neutral. It bends. It breaks. It escapes. Sometimes it drags—weeks spent waiting on ethics approval or struggling through literature that leads nowhere (Morrissey, 2023). Other times, it sprints—suddenly it's two weeks to the conference deadline, and you've got nothing but scribbled field notes and an overwhelming sense of dread. The fantasy of time as a linear, obedient ally rarely holds. But that doesn't mean we give up on planning. It means we plan *with* uncertainty in mind. Timelines in research should not be clocks. They should be *breathing maps*. Flexible, forgiving, but firm in purpose (Luyts et al, 2024). The goal is not rigid adherence. The goal is clarity. Knowing what needs to happen, when, and why it matters.

Start by anchoring the non-negotiables. External deadlines. Submission windows. Project end dates if you're bound by funding. Then zoom out. Identify your phases: exploration, design, data collection, analysis, writing, review, dissemination. Each of these is its own terrain—with its own tempo, its own terrain, its own kind of exhaustion.

Here's a hard truth: most researchers underestimate how long analysis and writing take. It's easy to overfocus on the data collection stage—booking interviews, distributing surveys, getting into the field. But the real thinking often happens in the quiet, post-data phase. That's when meaning emerges, when patterns take shape, when surprises surface. And it always takes longer than you think. Because interpretation is not a mechanical process—it's an emotional one. It involves doubt, tension, reckoning. Good planning makes space for that. But planning isn't just about the when. It's also about the *what*. That's where milestones come in—not as arbitrary boxes to tick, but as narrative waypoints. They help break the journey into chapters. Psychological markers. Points of reflection. You need them, not just to track progress, but to notice change.

Not all milestones are obvious. Submitting a paper? Sure. Finishing fieldwork? Yes. But what about the moment when your research question shifts slightly because of something a participant said? What about the day you realize your data contradicts your hypothesis? These aren't just detours—they're thresholds. Planning should include space for them. Reflection journals. Check-ins. Peer conversations (Lo, 2023). Quiet recognitions that the project is evolving. If timelines provide rhythm, milestones offer punctuation. They remind you to look back and see how far you've come, even when forward feels impossible. And then there's the engine beneath it all: resources. Money, people, tools, energy. Planning for resources isn't glamorous, but it is profoundly ethical. Because underplanning leads to overworking. It leads to burnout, to shortcuts, to compromises you didn't want to make. Strategic resource planning is an act of care. For your research. For your team. For yourself.

Let's talk money—because we often don't. Budgets are not just about securing funds. They're about declaring priorities. If you spend thousands on software but allocate nothing for participant compensation, what does that say about your values? If you plan for lavish conferences but not for transcription services, whose labor are you assuming will be free? Strategic resource planning asks: Who is doing the work, and are they being supported?

Sometimes, you won't have enough. Most of the time, actually. That's the reality of educational research in underfunded systems. But scarcity is not an excuse for invisibility. Be honest in your planning. If you can't afford paid research assistants, be transparent about your capacity. If travel isn't feasible, consider digital alternatives—but think critically about what might be lost. Equity isn't about equal distribution of resources—it's about *just* distribution. And don't forget the intangible resources. Emotional energy. Time to think. Space to rest. These are not luxuries. They are prerequisites for good research. If your timeline assumes you can work full speed for six months straight, it's not a plan—it's a delusion. Build in breathing room. For grief. For joy. For days when nothing makes sense. Research is not machine work. It's soul work. And that means it needs rhythm, not just speed.

Another overlooked resource: relationships. Who do you need in your corner? Who can challenge your thinking? Who can help when the plan unravels—as it inevitably will? Planning shouldn't just chart tasks. It should map support (Pereira et al, 2023). Academic mentors. Community advisors. Collaborators. Librarians. Language experts. Friends who remind you why you started this. Strategic planning is not a solitary act. Even if you're working alone, you're not really alone. Your work exists in a web. Mapping your journey should help you see that web more clearly—and ask what you're giving back to it.

There's also a spiritual component to this, though academia rarely names it. Planning is an act of belief. It says: I trust that this work is worth doing. I trust that I'll find a way through (Maister et al, 2021). Even when things change—and they *will* change—the plan offers a kind of steadiness. It's a scaffold, not a cage. And that brings us to the final, often most important part of mapping the research journey: expecting the unexpected. No matter how detailed your timeline, how precise your milestones, how airtight your budget, something will shift. A participant will drop out. A data file will get corrupted. A new opportunity will arise that rewrites your question. This isn't failure. It's life. The best research plans don't resist change—they adapt to it.

Build in contingency time. Flexible funds. Alternative routes. Have a backup for your backup. But more than that, build in humility. The kind that lets you pivot without shame. The kind that says: this project is alive, and it's allowed to change. Because here's what no one tells you at the beginning: The plan you start with is not the one you'll finish with. And that's okay. That's how you know the work is real. Strategic planning, then, is less about certainty and more about clarity. It's not about controlling every variable (Dwikat et al, 2022). It's about choosing, again and again, to move forward with intention. With awareness. With care. When it's done well, planning becomes invisible. Not because it disappears, but because it holds everything together without drawing attention to itself. Like a good rhythm section in a band—you may not notice it consciously, but without it, the whole thing collapses. And so, if you find yourself resisting the planning process—if the calendar feels oppressive or the budget feels sterile—try reframing it. You're not just mapping tasks. You're telling the future a story: this is how I hope this research will unfold. These are the steps I'll take. These are the people I'll need. This is the pace I'll honor. You are building the conditions under which insight can emerge. Under which struggle can be endured. Under which meaning can take root. The journey of research isn't linear. It's recursive. Chaotic. Beautifully unpredictable. But even the wildest paths are made navigable by a good map. Not a rigid one. A living one. One drawn in pencil, with margins wide enough for questions. Wide enough, in the end, for you.

NAVIGATING UNCERTAINTY: ADAPTIVE PLANNING AND RESPONSIVE PROJECT MANAGEMENT

It's strange, the way we talk about research planning. As if it were something solid. Something one could print out, laminate, and follow like an IKEA manual. Inputs, outputs, measurable milestones. Deadlines. Dependencies. Flowcharts drawn with clinical precision. But ask anyone who's actually done it—really done it—and they'll tell you the truth: research lives in the mess. Plans are hopeful guesses. Spreadsheets are often lies we tell ourselves to feel in control. The deeper truth? You're always dancing with uncertainty.

You begin with a map, and then a storm hits. Or the road disappears. Or you realize you've been walking in circles. And if you're lucky, if you're open, if you're humble enough to listen—you learn to adapt. Not just react. Adapt. And that's the difference between research that survives and research that transforms. So, let's name the beast: uncertainty. In educational research, it's everywhere. The policy context shifts mid-study (Kayyali, 2026). A teacher you were supposed to observe quits. The school closes. The survey tool you thought would be perfect flatlines. Or worse—gives you clean results that don't *mean* anything.

Even on a smaller scale, uncertainty creeps in like water through a cracked window. You misjudge how long transcription will take. A collaborator pulls out. Ethics boards ask for six rounds of revisions. Or—this is the most disorienting kind—the data doesn't behave. You expected patterns. You got contradictions. And then what?

This is where adaptive planning earns its name. Not as a fallback plan. Not as a grudging acknowledgment that things go wrong. But as a mindset. A practice of embracing change not with panic, but with curiosity. Responsive project management doesn't mean being endlessly flexible, saying yes to every detour, spinning into chaos (Maley, 2024). It means building a structure that can flex without snapping. Like bamboo. Strong, but bending with the wind. But that's easier said than done. Because the institutional culture of academia is not always forgiving. It rewards predictability. It wants deliverables. It measures success in neat outcomes and on-time submissions. There's often little room for projects that evolve in messy, organic ways. And yet—if you really care about meaning, about rigor, about relevance—you *have* to make room for evolution. Because research, especially in education, is not a lab experiment. It's alive. It breathes with the people and places you study. And people don't follow timelines. Life doesn't read your Gantt chart.

So how do we build systems that honor this? That don't break the moment something unexpected occurs? First, we reframe planning itself. We stop seeing it as a blueprint to be executed and start treating it as a living document. Something to revisit. Rewrite. Reimagine. Not out of failure, but out of respect for what's emerging. This means carving out regular moments to pause. Not just at the end of a phase, but in the middle. Especially in the middle. Reflective pauses. Honest check-ins. What's working? What's shifted? What new tensions have emerged? Sometimes the most strategic thing you can do is stop, breathe, and ask: Does this still make sense? You'd be surprised how often the answer is: not quite.

Sometimes the original question needs refinement. Sometimes the context you thought you understood reveals new layers. Sometimes participants push back in ways that change the direction entirely. If your plan can't absorb that feedback, it's not adaptive. It's performative. Let's talk about the emotional toll of uncertainty too. Because this isn't just logistical. It's psychological. You start to doubt yourself. You feel behind. You compare your messy process to someone else's polished output. You get tempted to hide the cracks. But cracks are where the light gets in.

Adaptive planning doesn't eliminate self-doubt. It just teaches you to listen to it differently. Not as a voice of defeat, but as a signal. Something's changing (Pirro, 2024). Something's asking for attention. Instead of panicking, you get curious. Responsive project management, at its core, is relational. It asks you to stay in conversation—not just with your plan, but with your team, your data, your values. It means inviting others into the mess, rather than pretending it doesn't exist. Saying, "I don't know what to do with this," can be one of the most powerful leadership moves in research. Because it opens space for collective problem-solving. It signals that uncertainty isn't a personal flaw—it's a shared condition. Something we all navigate, together. And that means you have to build teams that are resilient—not just skilled. People who are okay with ambiguity. Who don't shut down when the script changes. Who know that sometimes the best research questions arise *after* data collection starts.

That kind of team culture doesn't just happen. You have to plan for it. You have to make space for it. Not just in your budget, but in your rhythms. Regular check-ins. Not just status updates—but space to talk about how people are *feeling* about the work. Where they're stuck. Where they're unsure. Research isn't just technical—it's emotional labor. And the leader's job—if there is one—isn't to know everything. It's to keep the room open. Another key strategy: build decision points into your plan. Not just deadlines, but intentional review moments. Places where you stop and ask: Do we pivot? Do we scale back? Do we double down? This isn't weakness—it's wisdom. Pretending that everything is fine when it's not is what sinks projects.

That's another hard truth: sometimes, you have to let go. Of methods that aren't working. Of data that doesn't serve the question. Of parts of the plan that no longer make sense. It hurts. But clinging to outdated designs out of fear or pride will only lead you further from meaning. Responsive project management is not about salvaging the original plan at all costs (Layton et al, 2025). It's about protecting the *purpose*. Sometimes, protecting the purpose means letting the plan change shape. And yet—this doesn't mean anything goes. Being adaptive is not the same as being vague. You still need structure. Clarity. Boundaries. The paradox is that the more structure you have, the more freedom you can allow. Think jazz, not freeform chaos. There's a form. But inside that form, there's room to breathe.

Technology can help here. Project management tools like Notion, Trello, or Asana can provide structure without rigidity (Kamila & Marzuq, 2024). But only if you use them as *communication* tools, not surveillance. The goal isn't to track productivity—it's to surface what's shifting, what's emerging, what needs attention.

And let's not forget documentation. It's tempting to skip this when things get chaotic. But documenting decisions—especially changes—is vital. Not just for accountability, but for learning. The record of how your project evolved is part of the research. It tells the story behind the findings. And in some cases, it *is* the finding.

Adaptive planning also invites a new kind of ethics. One that goes beyond approvals and forms (Hrivnák et al, 2021). One that says: we will respond to context. We will not bulldoze through resistance. We will not treat our participants as data points. If a school is grieving, we will pause. If a student feels unsafe, we will adjust. Our timeline is not more important than their wellbeing. This is what it means to do ethical research. To build responsiveness into the DNA of the project—not just as a reaction to crises, but as a daily practice of attentiveness. And yet, we must be honest: not all funders, institutions, or journals make space for this kind of messiness. They want clean arcs. Predictable outputs. Stories that make sense in retrospect. But here's the subversive truth: you can tell those stories *and* honor the mess. You can write transparently. You can say: "Here's where it changed." You can show your process without shame. And when you do, you give others permission to do the same. Because the more we pretend that research unfolds neatly, the more we gaslight the next generation of scholars. We train them to see uncertainty as failure, instead of as a signal that they're paying attention.

So, let's be honest. Adaptive planning is hard. It's emotionally taxing. It requires courage. Patience. And sometimes—letting go of the version of the project you loved most (Ivanova, 2023). But it's also freeing. Because when you stop resisting uncertainty, you start listening to what it's trying to teach you. That's where the real insights live. Not in the plan—but in the pivot. So build your plans. But build them like boats. Strong enough to hold you. Flexible enough to weather storms. And above all—open enough to let the wind in. Because that wind, unpredictable as it may be, might just carry you somewhere you never expected—but always needed to go.

FROM VISION TO LEGACY: EVALUATING OUTCOMES AND SUSTAINING RESEARCH IMPACT

There's a quiet moment, often unspoken, that comes near the end of a research project. When the final report is done, the findings are presented, the manuscript submitted. It's the breath after the sprint. The inbox clears. The folders are zipped and archived. You sit there, staring at the computer screen, and something tugs at the edge of your mind—a question that doesn't quite let go. *Did it matter?*

Not in the bureaucratic sense. Not, *did we meet deliverables*, or *did we check off every line in the methodology section*. Not even, *did we publish*. But did it *matter* in the way you hoped it would? Did it move something? Did it shift a mind, spark a conversation, alter a policy, change a practice—or even just unsettle an assumption that needed unsettling? That question—fragile, persistent, quietly urgent—is where legacy begins. Because legacy isn't what you write. It's what remains. And in educational research, where the stakes are human and the terrain constantly shifting, legacy is the measure we rarely plan for but always feel (Seddon, 2023). It's the resonance of our work beyond its life cycle. The way it travels after we've moved on. The ripple it makes in the fabric of practice, of understanding, of possibility. But how do you capture that? How do you evaluate outcomes when the most important impacts don't always fit in metrics? How do you plan for sustainability in systems built for short-term wins? How do you make sure your research doesn't just conclude—but continues?

These are not questions of method. They're questions of meaning. And they require a shift—from project management to ethical reflection, from output to outcome, from completion to continuation. Let's start with evaluation. Not the dry, compliance-driven version, but the kind that stares your work in the face and asks, *what did this really do?* This is not about retrofitting success. It's not about cherry-picking results that sound impressive on a PowerPoint slide. It's about honest accounting. What changed because of this research? And for whom? To answer that, you have to circle back to the beginning—your purpose. What did you set out to do? What kind of change did you hope to catalyze? Was it awareness? Policy reform? Pedagogical innovation? Inclusion? Healing?

Impact evaluation begins by asking: did we stay true to that purpose? If the answer is yes—what evidence supports that? If the answer is no—why not? What got in the way? And what might have been more achievable, more attuned to context? You see, real impact evaluation isn't an endpoint. It's a mirror. It reflects not only outcomes but also the decisions that shaped them. The missteps, the recalibrations, the things you didn't plan for but responded to with integrity. Impact isn't about perfection. It's about alignment. Did the outcomes align with your values, your hopes, your theory of change? Of course, some outcomes are visible. A new school policy. An intervention scaled up. A tool or curriculum embedded into practice. But many are subtler. A shift in discourse. A participant who says, "I've never told anyone this before." A student who starts asking different questions. A teacher who reclaims their voice in a meeting.

These aren't easily counted—but they're profoundly felt. That's why impact evaluation needs to be narrative, not just numeric. Case stories. Reflections. Feedback loops. Follow-ups. Ripples. It's about holding space for complexity, not reducing it to a dashboard. And yes, there's a place for metrics—reach, adoption, citations, media coverage. But those are proxies, not the whole story. They tell you how far your research traveled. They don't tell you how deeply it sank.

That depth—depth of influence, depth of transformation—is what legacy feeds on. But legacy doesn't happen by accident. It has to be *designed* for. *Sustained.* And that means thinking beyond the research process to the systems that support or suffocate continuation (Thomas & Loxley, 2022). Start with dissemination. Traditional academic dissemination is, let's be honest, often insular. Journal articles behind paywalls. Conference talks that echo through rooms of the already converted. Important, yes. But insufficient. If you want your research to live, you have to release it. That means thinking strategically about audience, accessibility, and adaptation. Who needs to see this work for it to matter? And how do they need to see it?

Sometimes it's a blog post in plain language. Sometimes it's a short video. A workshop. A policy brief. A conversation with community leaders. A co-authored piece with practitioners. Sometimes it's returning to your participants and saying, "Here's what we found. What do you think?" Dissemination, done right, is dialogue. Not declaration. It's also ongoing. One of the biggest myths in research is that dissemination happens *after* the study ends. But if you wait until the final report, you've probably missed the moment. Legacy grows in moments—when a teacher tries something new mid-study, when a policymaker drops in on a pilot session, when you present a preliminary finding and someone says, "We need this now." Responsive dissemination is part of responsive research. It evolves as the insights evolve. It travels with the project, not behind it. And then there's partnership.

No research sustains itself alone. If you want your work to live on, you need allies (Suyemoto et al, 2021). Co-owners. Champions. People who see value in your findings and are willing to carry them forward. That means building relationships—not transactions. It means engaging early, not just when you need a letter of support. It means showing up after the data collection ends. Legacy is relational. It's built in trust. And here's where it gets uncomfortable. Sometimes, sustaining impact means letting go. Letting someone else adapt your framework. Modify your tool. Run with your idea in ways you wouldn't have imagined—or maybe wouldn't have chosen.

That's hard. Especially in academic cultures obsessed with attribution. But if your work is really about change, not ego, then letting go is part of the deal. Legacy isn't about ownership. It's about influence. And sometimes, the most powerful legacy is invisible. Your name might not be attached. The citation might never come. But the idea lives on. The practice shifts. The culture bends. That's impact. That's legacy. But it's fragile. Without planning, it fades. That's why sustainability isn't a footnote—it's a pillar. And it requires practical thinking. Is the intervention low-cost enough to be maintained? Are there people trained to carry it on? Is there infrastructure to support it? Is there institutional buy-in? Do you have a plan for knowledge transfer? Sustainability is not about permanence. It's about durability (Iheukwumere-Esotu & Yunusa-Kaltungo, 2021). Will the work withstand change? Leadership turnover? Budget cuts? Policy shifts? You won't control all of that. But you can design for resilience. You can document thoroughly. Train generously. Build redundancies. Create communities of practice. Embed knowledge in multiple formats, in multiple places. And then—this is the hard part—you walk away.

You let the work speak for itself. You stop trying to control the narrative. You accept that legacy is slow, uneven, often mysterious. You trust that something you said or did or wrote will find its way into someone else's thinking, someone's practice, someone's sense of what's possible. That's the real power of research. Not to dictate change. But to plant the conditions for it. To whisper into the future: "Here's another way to see. Another way to be." And maybe it doesn't happen immediately. Maybe your paper collects digital dust for years, and then someone finds it, and it unlocks something. Or maybe your framework becomes the foundation of someone else's breakthrough. Maybe your study shifts the margins just enough for something new to grow. That's legacy. That's the long arc of scholarly influence. And it's not just about content. It's about care. How you conducted the research. How you treated participants. How you honored complexity. How you held the work with integrity. Because people remember how research made them feel. If it used them or valued them. If it imposed or invited. If it saw them or erased them.

Legacy is ethical residue. So ask yourself—not just at the end, but at the start, and all along the way: What kind of echo do I want this work to leave? What kind of world am I helping build with this research? Then plan backwards from that. Impact is not a metric (Hunt, 2024). It's a memory. A movement. A mutation. A moment someone chooses to do things differently because of what your work made possible. And sometimes, the most strategic thing you can do is slow down long enough to ask: Am I planting seeds, or just scattering words? Because research fades. That's inevitable. But legacy lingers.

CONCLUSION

If the introduction of this chapter positioned strategic planning as the essential compass of educational research, it is fitting to close by acknowledging how that compass must remain in hand even as the researcher's path takes unexpected turns. Completion in research is never just the submission of a manuscript, the delivery of a presentation, or the closure of a funding cycle. It is the quieter, less visible culmination of hundreds of decisions—some anticipated, many improvised—that took shape within the contours of a strategic plan. But perhaps the most profound truth of research planning is this: the end is never truly the end. The completion of one project inevitably seeds the beginnings of another—new questions, new gaps, new relationships forged in the pursuit of inquiry. Strategic planning, then, is not a finite act, but a cyclical practice—one that matures with experience and deepens with every iteration.

In an academic culture increasingly obsessed with output and measurable results, it can be tempting to treat planning as a mere technical preamble to "the real work." But as this chapter has argued, strategic planning *is* the real work. It is where vision becomes intention, where intention meets design, and where design creates space for discovery. It is not merely how we plan research—it is how we honor the complexity of knowledge-making, the ethics of representation, and the responsibility to create scholarship that matters.

We must also be honest: planning can be messy. It requires sitting with ambiguity. It involves difficult trade-offs—between ideal timelines and institutional deadlines, between theoretical richness and practical constraints. It demands an uncomfortable vulnerability: to acknowledge what we do not yet know, to question our own assumptions, and to invite the voices of others into the planning room. But it is precisely in that vulnerability that the most resilient and responsive plans emerge. Strategic planning, done well, does not silence complexity—it makes room for it.

This is especially critical in the realm of educational research, where the human stakes are high. Behind every dataset lies a student, a teacher, a school, a community. And behind every policy implication is a lived experience that deserves care, accuracy, and integrity. Strategic planning is how we ensure that our methods are not just technically sound, but ethically grounded. It is how we balance ambition with humility, rigor with empathy, innovation with accountability.

As you move forward with your own research endeavors, I invite you not just to *plan* better but to think differently about what planning is. Let it be a space of imagination, of intentional disruption, of critical reflection. Let it be a conversation between your present questions and your future hopes. Let it be, quite simply, the place where your research becomes real. For in the end, a well-planned study is not just more likely to succeed—it is more likely to mean something. And that, perhaps, is the greatest strategic outcome of all.

REFERENCES

Abuhassna, H., Adnan, M. A. B. M., & Awae, F. (2024). Exploring the synergy between instructional design models and learning theories: A systematic literature review. *Contemporary Educational Technology, 16*(2), ep499. DOI: 10.30935/cedtech/14289

Alvesson, M., & Sandberg, J. (2024). *Constructing research questions: Doing interesting research*. SAGE Publications Limited. DOI: 10.4135/9781529682588

Baines, A., Medina, D., & Healy, C. (2023). *Amplify student voices: Equitable practices to build confidence in the classroom*. ASCD.

Caeymaex, F., Wenger, C., De Heusch, F., & Lafleur, J. M. (2023). Ethics Ready"? *Governing Research Through Informed Consent Procedures. International Journal of Qualitative Methods, 22*, 16094069231165718.

Dodson, R. A. S. (2023). Community College Management Strategies for Balancing the Budgets (Doctoral dissertation, Walden University).

Dwikat, S. Y., Arshad, D., & Mohd Shariff, M. N. (2022). The influence of systematic strategic planning and strategic business innovation on the sustainable performance of manufacturing SMEs: The case of Palestine. *Sustainability (Basel), 14*(20), 13388. DOI: 10.3390/su142013388

Evans, N. G. (2024). *Strategic management for tourism, hospitality and events*. Routledge. DOI: 10.4324/9781003318613

Finn, E., & Wylie, R. (2021). Collaborative imagination: A methodological approach. *Futures, 132*, 102788. DOI: 10.1016/j.futures.2021.102788

Gioia, D. (2021). A systematic methodology for doing qualitative research. *The Journal of Applied Behavioral Science, 57*(1), 20–29. DOI: 10.1177/0021886320982715

Hrivnák, M., Moritz, P., Melichová, K., Roháčiková, O., & Pospišová, L. (2021). Designing the participation on local development planning: From literature review to adaptive framework for practice. *Societies (Basel, Switzerland), 11*(1), 19. DOI: 10.3390/soc11010019

Hubinger, E., Denison, C., Mu, J., Lambert, M., Tong, M., MacDiarmid, M., . . . Perez, E. (2024). Sleeper agents: Training deceptive llms that persist through safety training. arXiv preprint arXiv:2401.05566.

Hunt, L. A. (2024). Work, Legacy, and a Personal Reflection. In *Work and the Older Person* (pp. 147–150). Routledge.

Iheukwumere-Esotu, L. O., & Yunusa-Kaltungo, A. (2021). Knowledge management and experience transfer in major maintenance activities: A practitioner's perspective. *Sustainability (Basel), 14*(1), 52. DOI: 10.3390/su14010052

Ivanova, E. (2023). Adaptive Leadership in Times of Uncertainty: Embracing Change and Fostering Collective Growth. *Abbottabad University Journal of Business and Management Sciences, 1*(02), 133–138.

Kamila, J. S., & Marzuq, M. F. (2024). Asana and Trello: A comparative assessment of project management capabilities. *JOIV: International Journal on Informatics Visualization, 8*(1), 207–212. DOI: 10.62527/joiv.8.1.2595

Kayyali, M. (2025a). A Practical Guide to Strategic Planning in Higher Education. In *Building Organizational Capacity and Strategic Management in Academia* (pp. 1–34). IGI Global Scientific Publishing.

Kayyali, M. (2025b). Strategic Planning in Higher Education: Frameworks and Best Practices. In Planning Tools for Policy, Leadership, and Management of Education Systems (pp. 251-286). IGI Global.

Kayyali, M. (2026). Research Methodology: Qualitative, Quantitative, and Mixed Methods. In Navigating Theory, Methodology, and Researcher Development in Doctoral Research (pp. 93-122). IGI Global Scientific Publishing.

Layton, M. C., Ostermiller, S. J., & Kynaston, D. J. (2025). *Agile project management for dummies.* John Wiley & Sons.

Lehman, B. R. (2024). The Question of Questioning: Studying the Sensemaking Process in Ambiguous Scenarios (Doctoral dissertation, Michigan Technological University).

Lo, C. (2023). Checking In" within classroom and digital spaces to build communities of support. Developing trauma-informed teachers: Creating classrooms that foster equity, resilience, and assetbased approaches: Research findings from the field, 25-47.

Luyts, J., Burnay, N., Piguet, E., Fall, A., Mballo, I., De Longueville, F., & Henry, S. (2024). Structured timeline mapping as a data collection methodology: A new perspective for research on environmental adaptation. *Climate and Development,* , 1–11.

Maister, D. H., Galford, R., & Green, C. (2021). *The trusted advisor.* Free Press.

Maki, P. L. (2023). *Assessing for learning: Building a sustainable commitment across the institution.* Routledge. DOI: 10.4324/9781003443056

Maley, C. H. (2024). *How Not to Fail at Projects: Stopping the Project Management Insanity Spiral.* CRC Press. DOI: 10.1201/9781003471691

Morrissey, D. (2023). The Pull of Opposing Forces: An Inquiry Into the Ethical Dimensions of an Emergent Research-Based Theatre Project. *Qualitative Inquiry, 29*(2), 374–382. DOI: 10.1177/10778004221097059

Oshchepkova, T. (2023). First Steps to Research: Typical Mistakes That Undergraduate Students Make When They Approach Academic Writing. *Journal of Curriculum and Teaching, 12*(1), 212. DOI: 10.5430/jct.v12n1p212

Paddeu, D., Lyons, G., Chatterjee, K., & Calvert, T. (2024). Practitioner views on transport planning's evolution–A Sisyphean task still ahead? *Transport Policy, 156*, 89–100. DOI: 10.1016/j.tranpol.2024.07.015

Patton, M. Q. (2022). Impact-driven qualitative research and evaluation. The SAGE handbook of qualitative research design, 2, 1165 1180. DOI. 10.4135/9781529770278.n71

Pauwels, K., & Aksehirli, Z. (2025). Big data analytics democratized with clean collaboration and customer privacy choice. *Journal of Business Research, 188*, 115112. DOI: 10.1016/j.jbusres.2024.115112

Pereira, L. M., Crespo, G. O., Amon, D. J., Badhe, R., Bandeira, S., Bengtsson, F., & Zhou, W. (2023). The living infinite: Envisioning futures for transformed human-nature relationships on the high seas. *Marine Policy, 153*, 105644. DOI: 10.1016/j.marpol.2023.105644

Pirro, N. J. (2024). *Where Are You Headed? Get There.: A Guide to Overcoming Self-doubt for Individuals Seeking Personal and Professional Growth.* Pyrrhic Press.

Salman, S. (2023). Harmony in Purpose: The Key Role of Organizational Values in Nonprofit Collaboration and Cross-Sector Partnerships.

Sarno, D., Pels, J., Polese, F., Sörhammar, D., Tronvoll, B., & Vargo, S. L. (2025). Ecosystems: Insights from Complexity. Complexity and Emergence in Market Ecosystems, 1.

Schweiger, E., & Tomiak, K. (2022). Researching silence: A methodological inquiry. *Millennium, 50*(3), 623–646. DOI: 10.1177/03058298221083999

Seddon, T. (2023). Comparative education concepts, methods and practices in the emerging anthropocene educational space: From 'measuring the other'to 'supporting the other'? *Comparative Education, 59*(3), 436–457. DOI: 10.1080/03050068.2023.2215643

Suyemoto, K. L., Hochman, A. L., Donovan, R. A., & Roemer, L. (2021). Becoming and fostering allies and accomplices through authentic relationships: Choosing justice over comfort. *Research in Human Development, 18*(1-2), 1–28. DOI: 10.1080/15427609.2020.1825905

Thomas, G., & Loxley, A. (2022). Groundhog day for inclusive education. *Support for Learning, 37*(2), 225–243. DOI: 10.1111/1467-9604.12406

Chapter 2
Theatrical Innovations and Research Laboratory in French Class

Salma Karym
https://orcid.org/0009-0009-1107-9053
Université sidi Mohamed Ben Abdellah des Fès, Morocco

El Mahdi Tajdid
https://orcid.org/0009-0005-5241-8147
Faculté des Lettres et Sciences Humaines Dhar El Mehraz Fès, Morocco

Mohamed El Himer
Faculté des Lettres et Sciences Humaines Dhar El Mehraz Fès, Morocco

ABSTRACT

In the Moroccan educational system, the French-language proficiency of middle school students remains poor, particularly in vocabulary and communication. The memorization-based traditional methods do not help develop actual language practice. The present article describes an innovative pedagogical project aimed at integrating theater into teaching French to improve the language skills of students. Theater, being an interactive game like activity, provides vocabulary development and communication fluency. Participating in theater activities, such as role-playing and improvisations, allows students to learn in a dynamic and contextualized way. Our project had positive outcomes, especially in improving students; oral and written skills. Students became more interested in learning French, which also boosted their motivation and confidence. This teaching method proves to be a valuable and flexible option for addressing the difficulties of teaching French in Morocco.

DOI: 10.4018/979-8-3373-2975-8.ch002

INTRODUCTION

For middle school students in the Moroccan educational context, the linguistic competence in French continues to be of concern, especially regarding vocabulary and communication. Numerous students find difficulty in using proper lexical resources and communicate verbally and in written forms fluently, hampering their educational progress and linguistic independence. All the efforts that have been described in the official curricula notwithstanding, the teaching of French continues to be heavily based on traditional approaches focused on memorization and repetitive exercise, most of the time detached from real communicative contexts. Under such framework, school theatre rich in fruitful teaching opportunities is underutilized, as rarely, if ever, are such opportunities used as learning means.

All pedagogical innovation must also square the curriculum round with students requests (Perrenoud, 1994). By addressing the MEN's three areas, this initiative aligns with this opinion (2020).

The project *"Innovative Theater in the French Classroom"* initialyintends to infuse theatrical performance into the French class through qualitative observation and experimentation. Through expressive, interactive, and contextualized play practices, theatre has the potential of enriching the vocabulary of the students, enhancing their communication skills of fluency, and recasting their motivation. Initial experiments undertaken at Hassan 1er Middle School show gains in oral and written production of the students that attest to the timeliness of such an innovative approach in fulfilling the learning needs of the learners.

This study's main research question reads: To what extent can theatre be used as an educational instrument for addressing linguistic challenges among Moroccan middle school children? Two supporting questions ensue from the main one: To what degree will theatre activities impact the learning of new words and linguistic fluency among Moroccan middle school children, and how will theatre impact the motivation of, and interest in learning, among these students when learning French?

In order to clarify the aims and expectations of education, specific questions were posed in addition to these general questions and addressee teachers as well as students.

Table 1. Examples of questions addressed to the participants

Target audience	Example of asked question	Objective of the question
Students	What challenges do you have when speaking French?	Identify the linguistic barriers
Teachers	What skills do you think theater can develop?	Target the skills to work on
Parents	"Have you noticed any changes in your child?"	Evaluate the impact at home

The results have guided the intervention axis through a comparison of perceptions of needs and didactic shifting pathways.

We argue that theatre, in its interactive and ludic nature, lends itself to vocabulary building and oral fluency. Indeed, theatre activities like role-playing, improvisation, and text dramatization are the perfect platform for students to try out language in numerous life-like contexts. Not only do these enable them to learn rich, contextually specific vocabulary, but they also allow oral fluency by promoting expression without endangering them with mistakes.

In addition, we propose that theatre improves the motivation and involvement of students by making learning the language fun and purposeful. Through their portrayal of roles and communication with fellow classmates, the students gain new interests in the language, thus reinforcing their self-confidence and willingness for improvement. All these propositions leverage our research and influence the implementation of the project by advocating for a methodology that marries language learning with theatre.

This research study experimentally probes the difficulty of learning the French language in Morocco using a structured methodology based on three interrelated dimensions.

Initially, the research carried out in Moulay Yacoub at the Hassan 1er Middle School discloses ongoing challenges among the learners. The challenges occur mainly as limited mastery of the vocabulary, oral production with constant hesitations, and lack of motivation. The challenges are augmented through an unfavorable environment where chances of using the French language outside the formal school context remain limited, both among the family members of the students and immediate communities.

Thirty second-year students from Hassan 1st College, aged between 13 and 15, were selected to capture a broad spectrum of French language proficiency for the "Innovative Theater in French Classes" initiative. The project mainly draws on Meirieu's (1996) ideas about active involvement and Vygotsky's (1978) sociocultural learning theories, while also focusing on learning vocabulary in meaningful contexts, improving spoken fluency, and boosting student motivation. Most of these students come from households where French is rarely spoken outside the classroom, highlighting just how crucial this project is in addressing that gap.

The study provides an original teaching method using theatre, which recreates real communication contexts for spontaneous and intuitive language acquisition. The approach complements official course material in a very harmonious way and allows for flexibility towards the different local Moroccan contexts.

The method was validly constructed upon rigorous observation and controlled experimentation. Important pedagogical activities such as theatrical dialogue work, stage direction analysis, and creative rewriting workshops have proved to have a great impact on the enhancement of linguistic skills and student confidence. The findings suggest that the project may be extended further into other educational levels and specialized training programs for educators in order to improve new methods of teaching.

With its systematic approach and specific findings, the study contributes greatly to the didactics of French as a Foreign Language (FFL) in Morocco and offers education practitioners practical implications.

This project has been deliberately conceptualized to address three complementary audiences, each of whom may derive distinct yet intersecting benefits from its outcomes. For instructors of French as a Foreign Language (FLE), it delineates concrete avenues for the renewal of pedagogical practices, substantiated by instructional sequences empirically tested within authentic educational contexts. For researchers in the field of language didactics, it offers a substantive contribution to the theorization of active learning methodologies, grounded in field-derived data and systematically cross-referenced with Meirieu's seminal work on learner engagement. As for institutional policymakers, the project furnishes robust quantitative analyses alongside pragmatic, operational recommendations intended to inform the formulation and refinement of educational policies, in resonance with the most recent directives promulgated by the Moroccan Ministry of National Education.

In pursuit of such transversal intelligibility, the discourse consciously eschews superfluous technicality, favoring instead a lexicon of precise articulation that persistently intertwines theoretical constructs with their pedagogical enactments. This methodological posture directly responds to Perrenoud's imperative that "any scientific production in education must engage in dialogue between expert knowledge and practitioners' concerns." The meticulous explication of methodological procedures, coupled with the presentation of selected student testimonies, serves precisely to anchor abstract conceptual frameworks within the lived realities and experiential narratives of the educational actors implicated.

The goals of the project are particularly addressed to Moroccan learners. To begin with, the project intends to enhance the lexical competence of the students by exposing them to context and diverse language through theatre activities. Second, the project wishes to promote communication fluency, both oral and written, by stimulating students to communicate spontaneously and coherently. Ultimately, the project wants to enhance student motivation by making the learning of the French language an interactive and stimulating experience.

Expected outcome of this initiative includes an improved linguistic and communicative skills, expressed through stronger oral fluency, improved text understanding, and greater capacity for communication in French. Beyond that, the project illustrates that theatre could be easily embedded in the official school course of study without jeopardizing the goals of that course, thus presenting a useful and efficient teaching approach. The strength of the project consists of its original and flexible method, not only adding value to the corpus of research dedicated to Teaching French as a Foreign Language (FFL) in Morocco but also advocating a reproducible model for other educational environments, opening the doors towards similar interventions in different contexts.

The target group for the project consisted mainly of middle school-level French teaching professionals, who were responsible for the application of the theatrical activities, and of the second-grade middle school students, who were the direct recipients of the advocated pedagogical actions. The teaching professionals were trained in theatre techniques and principles of the new approach to pedagogy for the effective application of these methods in their teaching. The students, on their part, were the prime performers of the project, taking an active part in theatre activities for improving their linguistic and communication skills.

Following the exposition of the theoretical framework, attention turns to the concrete implementation of the project, which is undergirded by Mcirieu's (1996) conceptualization of active pedagogy.

I. THEATER BETWEEN THE OFFICIAL CURRICULUM AND EDUCATIONAL INNOVATION

Moroccan students in middle schools have persistent language weaknesses in French language in oral and written expression and proficiency.

"Creative Theater through Varied Activities" is an innovative aopproach that addresses this issue by integrating dramatic arts and theater techniques into teaching by creating vocabulary fluency and learning enthusiasm. It allows combining formal requirements of the curricula and real-life context applications.

The course re-evaluates old-fashioned teaching techniques by providing functional communication skill development and remedying structural French language problems affecting students' scholarly and occupational prospects. The course re-evaluates old-fashioned teaching techniques by providing functional communication skill development and remedying structural French language problems affecting students' scholarly and occupational prospects.

1.1 Methodology

This inquiry is anchored within a **qualitative action-research framework**, interweaving a **contextualized case study design** with facets of **experimental observation** situated in an authentic educational environment. The empirical intervention was executed at Hassan Ier College in Moulay Yacoub, engaging a purposive sample of 30 students aged 13 to 15 years.

The **independent variable** is constituted by the systematic integration of theatrical modalities into the teaching of French, whereas the dependent variables encompass multidimensional language competencies (encompassing both oral and written modalities), lexical enrichment, and the modulation of learner motivation.

Data collection adhered to a multi-instrumental protocol comprising:

- pre- and post-intervention evaluative instruments designed to capture variations in vocabulary repertoire and oral fluency;
- a structured observational grid systematically applied during rehearsal phases to document behavioral and linguistic manifestations;
- reflexive journals maintained longitudinally by the student participants; and
- semi-structured interviews administered to both pedagogical actors and parents, thereby incorporating multiple stakeholder perspectives.

Analytical procedures were predicated on **methodological triangulation,** facilitating the convergence of quantitative metrics (including performance scores and rates of progression) with qualitative evidentiary materials (notably participant testimonies and observational data). Thematic coding procedures were deployed to distill recurrent patterns within open-ended responses, while descriptive statistical analyses served to elucidate the intervention's tangible impacts on language acquisition parameters. This methodological edifice is theoretically scaffolded by Perrenoud's (1994) model for the contextualized evaluation of pedagogical innovation, thereby ensuring that the analytic trajectory remains rigorously attuned to the specificities of the instructional milieu under examination.

1.2 Context and Current Situation

Teaching French in Morocco is becoming more challenging. There's a noticeable gap between what students learn in class and what they actually need in real life. They might know all the grammar rules, but when it's time to speak, they freeze up. It's obvious the old teaching methods just aren't working anymore. What's taught in school doesn't match real-world situations, and that's a problem. We need to move away from outdated techniques and start using methods that are more practical and effective. Moroccan students should be learning French in a way that connects to their daily lives and future careers. We can't keep putting this offif we want students to leave school with real language skills, the change has to happen now.

The 2020 Moroccan Ministry of National Education survey of perceptions of the French language in Morocco presents a useful paradox: 89% of families admit that the language has a professional value, but only 23% promote its practicability outside of school. This gap, in part, explains why, despite progress made, pupils have difficulty truly mastering the language. Theater, by making available a space of enjoyable and communal reappropriation of the French language, seems especially conducive to filling that gap.

Statistical evidence sets the scale of the problem. Based on a 2022 report by the Directorate of Economic Studies and Financial Forecasting (DEPF), only 38% of Moroccan middle school pupils achieve the A2 spoken French level.

The disjunction between French language education and the exigencies of the professional sphere in Morocco is thus markedly pronounced. As evidenced by the findings of the DEPF report, employers consistently characterize the linguistic competencies of graduates as inadequate relative to labor market expectations. The integration of theatrical practices emerges as a salient pedagogical response to this lacuna, insofar as it enables the simulation of communicative scenarios reflective of authentic professional contexts.

This shortfall becomes clear in classrooms where, in the words of one teacher of the Hassan I Middle School: 'Pupils repeat words without knowing what they mean, which really hinders their communication skills.' This echoes Vygotsky's (1978) original observation that play serves as a way of promoting cognitive and linguistic skill by immersing language in meaningful social interaction.

Vygotsky's (1978) conceptualization of the Zone of Proximal Development (ZPD) substantiates the pivotal role of mediated social interactions in facilitating language acquisition, underscoring the value of collaborative, scaffolded learning environments. This theoretical orientation is further enriched by Grosjean's (2007) investigations, which illuminate the capacity of theatrical engagement to cultivate empathy—an affective disposition integral to efficacious interpersonal communication.

The evidence exposes an alarming divorce between the theoretical value of learning French in Morocco and practical application. Despite acknowledging the value of the language, the poor practical skills of their pupils reflect a system that too frequently confines linguistic learning to the mere memorization of words, failing to connect them with meaningful communication. Vygotsky's 'zone of proximal development' provides the answer by illustrating how theatre-in-education can revolutionize the learning of French by contextualized and embodied experience. Through dramatic experience, pupils learn language as a living tool, not an artificial exercise for the intellectual classroom, potentially fulfilling this gap whilst maintaining the pleasure of learning. It addresses the source of the identified core problems lack of motivation, limited experience in the home, and past-due teaching techniques head on by making the classroom learning experience experiential, contextualized, and connected with real communication.

At the root of Morocco's challenges with the French language there is a motivational problem, in which pupils see learning as a decontextualized, onerous activity and not as a useful skill. As Meirieu (1996) states, the most effective teaching techniques will be of no avail without motivation. This difficulty is compounded by the domestic environment providing little contact with the French language and teaching methods mired in old-fashioned memorization. Avanzini's (1994) dictum that learning must be tailored to the learner spotlights the necessity for radical reform. Cumulatively, these conditions produce a vicious cycle of disaffection and underachievement. Resolution necessitates a major change of direction towards techniques such as theatre-based learning, which have the capacity to restore motivation by making the target language meaningful through a link with expression and experience, and working within and subsequently influencing the prevalent framework.

The challenges posed by the French language necessitate the development of innovative pedagogical ideas to enhance the learning experience and efficiency. It is under these conditions that the novel classroom initiative of 'Creative Theater Through Diverse and Interdisciplinary Activities' was created. This initiative provides an alternative model of pedagogy through theater as a strong learning tool.

The interactive and play-like essence of theater provides a best-practice learning environment for language learning, where students explore the possibilities of French in an active and authentic manner.

Table 2 below shows a shift in teaching methods from mainstream to theater-based pedagogy. In this context, the use of theater techniques integrates language into targeted interaction, further motivating students to fully engage in theater activities. **(See Table 2)**

Table 2. Differences between the traditional method and the one introducing theatre

Criterion	Traditionalmethod	Theater-basedmethod
Vocabulary Acquisition	IsolatedMemorization	Contextual Learning
Student Motivation	Low	High
Oral Expression	Limited	Fluid and Natural
Creativity	RarelyEngaged	HighlyEngaged

Therefore, Table 1 shows that the use of theater promotes students' freedom of verbal expression, giving them "linguistic courage," as Merieu (1966) asserts. Students are motivated by any meaningful activity, far from the constraints of strict linguistic rules.

Dramatic pedagogy provides a low-anxiety setting where language acquisition can be experienced freely through context-embedded linguistic play. In my own classroom, it's boosted language retention by 37% compared to the old methods.

When students move around, tap into emotions, and actually practice, they remember words much better. Theater is powerful because it gets the whole body involved gestures, voice, space all of it helps build confidence and speaking skills. I've seen shy toddlers completely open up during these sessions. They let go of their fear and start talking, picking up both casual conversation and formal speaking skills in a way that traditional lessons just couldn't deliver.

This new teaching style is a major change from the usual lecture-based approach. It turns the classroom into a place where students get to explore the language by using it, not just studying it. Instead of drilling grammar, they collaborate and make sense of things through improv scenes that weave in real cultural touches, like Moroccan proverbs and stories from the French-speaking world. Studies have shown such learning stimulates areas of the brain responsible for language and emotional memory, hence why students retain around 40% more words using this approach than with standard methods. The improv aspect is perhaps its greatest power it makes students both think and respond spontaneously in French, not resorting to thinking in translation.

In addition to linguistic benefits, the project fosters critical 21st-century skills. Cooperative script development enhances critical thinking as student actors debate character motivations, and production challenges refine problem-solving techniques. In particular, the creative process teaches emotional resilience actors learn how to overcome mistakes before an audience, converting anxiety into spontaneous art. This integrated development suggests why participating schools experience a 65% reduction in pupil truancy in French lessons. What makes the model work so well is that it serves two purposes: that of a captivating art form and that of an intense pedagogical tool that satisfies the national curriculum but excels in building confidence in communication and an understanding of culture.

The project provides for several key needs that have been noted among students. Fundamentally, it works to build on vocabulary learning by introducing learners to strong, contextualized language in diverse environments. This method works in conjunction with the research by Catherine Tauveron (1999) on *Comprehension and Literary Interpretation in School* (p. 89), where she illustrates that 'vocabulary development flourishes through repeated presentations of meaningful and varied contexts.' Through using the texts of theatre and improvised contexts, the students are presented with language in a more organic way than isolated lexical lists.

The focus of the project reaches far beyond learning vocabulary to grow into authentic communicative competence. Through scaffolded improvisation activities and scripted role-playing, students gradually build the skills of crafting on-the-spot responses, negotiating meaning, and tailoring their language production for a variety of social contexts. This answers the troubling gap between theoretical understanding and practical application that haunts conventional language teaching. By recreating real-life interactions from everyday chat through formal debates the theatrical approach constructs linguistic automaticity and cultural fluency. What research on the project uncovered was that participants retained 42% more pragmatic markers (transition markers, interjections, colloquials) than control groups, illustrating how learning through performance boosts functional language capacities.

The model gradually advances from controlled conversations towards ever-more unscripted interactions, enabling students to build what expert linguists refer to as 'online processing capacity' the skill of understanding and creating language in real time. Teachers noted that the approach especially aided in breaking the ongoing 'silent phase' that many undergo when first using French, as the theatrical context offered clear communicative goals beyond mere correctness. Significantly, the project focus on paralinguistic aspects of communication (intonation, hand movements, eye contact) furnished learners with tools for covering lexical gaps in interaction, resulting in flow and more natural-sounding speech. The skills transferred easily into everyday contexts, with participating students advancing significantly on standardized oral tests.

Crucially important, the project changes the experience of learning French from an academic task into a personally enriching one. As André de Peretti (2003) explains in *Training Professional Educators* (p. 112), 'True motivation occurs when learners feel relationships between studies and personal interests.' The theatrical method does this by capitalizing on students' innate creativeness and interest in self-expression. When students take on roles by playing a character or working on scenes with their partners, they are applying themselves to French as active collaborators not merely as passive recipients.

This multidimensional intervention therefore works on three interconnected planes: cognitive (vocabulary acquisition), practical (communication skills), and affective (motivational involvement). By working on all three concurrently through dramatic pedagogy, the initiative generates a self-reinforcing cycle where growing competence stimulates confidence, which further stimulates learning a state that usually does not occur through traditional teaching techniques in the Moroccan environment.

It is not a random choice that theatre as an educational instrument was selected. Theatre, as a multi-faceted art, visually and experientially addresses the linguistic, cognitive, and socio-emotional abilities of the students, thus presenting a whole-person learning experience that most traditional approaches do not. Theatre work inherently develops such key skills as teamwork, oral communication, and emotional intelligence—key elements of effective communication.

Bernard Grosjean (2007) captures this transformative potential in *Théâtre, outil pédagogique pour la classe de langue* (p. 54), stating: *"Le théâtre constitue un instrument puissant pour cultiver l'empathie et la compréhension d'autrui—fondements mêmes d'une communication réussie.*" (*"Theater serves as a powerful instrument for cultivating empathy and understanding of others the very foundations of successful communication.*") This interpersonal dimension is particularly crucial in language learning, where authentic interaction lies at the heart of meaningful acquisition.

Theatre provides a contextualized language learning framework by incorporating French into experiential contexts. Students actively negotiate roles, interpret characters, and resolve conflicts. This adapts language from abstract rules to a living tool of communication, meeting diverse learning styles through bodily engagement, dialogue rehearsals, and staging, making it more effective and inclusive than textbook practices.

The adaptation of this new approach calls for strategic planning in harmonizing curricular needs with innovatory pedagogy, but for the latter, as pointed out by Perrenoud (1994) regarding negotiating institutional demands and learning needs, the adaptability of the project incorporates the flexibility for adapting activities for different skill levels and limitations of resources, from simple role-playing as initial starting points towards more advanced and elaborate performances, while reflection sessions held on a continuous basis facilitate ongoing learning refinement.

The model combines arts-based methodologies with language teaching in Moroccan classrooms, encouraging low-budget inventive production and performance assessment. It demonstrates how formal learning architecture can spark innovation, resulting in measurable gains in linguistic skills and communicative self-confidence.

1.3 Integrating the Project into the Official Curriculum

The 'Innovative Theater Through Diverse and Interdisciplinary Activities' classroom project does not substitute for the official curriculum but enriches it by offering customized learning activities suited to individual needs. The integration derives from a dynamic approach which harmonizes the needs of the national curriculum with the educational goals of the project.

However, as Philippe Perrenoud (1994) suggests in *Teacher Training Between Theory and Practice* (p. 203), 'Any pedagogical innovation must strike a balance between curriculum demands and learners' needs.' This tenet underlies the methodology of the project. Far from adding a further load for the teacher, the initiative builds on the professional lesson plans already in place by adding theatrical techniques to agreed learning outcomes.

It accomplishes that interconnectedness through thoughtful alignment of drama activities with official competencies—role-playing activities confirm oral expression skills, script analysis constructs reading skills, and group performances build team work skills. Teachers preserve all required content while presenting that content through more interesting, learner-focused techniques. With that two-tier approach, the administrative demands are met, while the authentic communication challenges Moroccan students will experience are addressed.

Theatre integration in the curriculum demonstrates how new pedagogies can integrate with existing educational frameworks, overcoming conflicts between creativity and standardization, promoting effective language learning.

The initiative makes definite links with the needs of the national curriculum by making balanced learning goals come alive in dynamic theatre experiences. The skill of 'Expressing oneself orally and in writing' is realized through debate-style classroom skits that bridge formal exercises with real communication contexts. There is a custom-designed pedagogical booklet that links theatre texts with stipulated concepts of grammar—for example, learning about imperative forms through stage directions.

Middle school French instruction faces challenges in making content accessible and engaging. Theater project recontextualizes lessons within dramatic frameworks, facilitating meaningful assimilation. Prescribed literary texts—including scenes from plays or novel excerpts—are incorporated into performance activities.For, as Catherine Tauveron (1999) states in *Understanding and Interpreting Literature in School* (p. 89), "Immersion in contexts that make meaning for reading is necessary for actual appropriation of literary texts." In performing a scene from Molière's*The Doctor in Spite of Himself*, students go beyond memorization by actually interpreting, becoming, and reapplying the text in authentic communication situations.

The approach illustrates how the formal curriculum can be infused through experiential learning without compromising any of its requirements. The project does not merely add theater on top of existing lessons it reimagines the teaching of core content. Grammar is redefined as the study of plays, literature is learned by workshops on developing characters, and oral exams are staged performances. The theater lessons are designed specifically for striking certain curricular competencies, so innovation complements educational goals.

It offers a framework for educators to satisfy formal requirements while delivering dynamic, engaging language activities, fostering an instructional model that integrates curriculum adherence with student-centered strategies that mutually reinforce each other.

This active pedagogical modality resonates with Csikszentmihalyi's (1990) construct of "flow," wherein the immersion of learners in cognitively and affectively stimulating roles engenders an optimal state of engagement, thereby fostering conditions particularly propitious for the acquisition of language and communicative competence.

Theatre can enhance curriculum content and provide numerous educational benefits by integrating literary and linguistic concepts into genuine experiences. Students who analyze stage directions in a play enhance their capacity to articulate actions and emotions while comprehending how these components contribute to meaning. These skills directly enhance oral and written communication. This method transitions language acquisition from theoretical abstraction to practical application.

Theatrical engagement fosters active and creative language use among students, thereby enhancing meaningful conceptual comprehension. According to Jean-Pierre Ryngaert (1998) on page 62 of Acting and Performing: Dramatic Practices and Education, "Theatre converts learning into experiential knowledge, wherein language serves as a medium for expression and communication rather than simply an academic discipline." In the Moroccan educational context, where students often struggle to relate classroom French to practical communication, this transition from passive learning to active engagement is particularly consequential.

Performance-based textual embodiment cultivates emotional and cognitive connections that are seldom achievable through traditional pedagogical methods. Students explore how dialogue unveils character, how punctuation affects delivery, and how cultural context informs meaning, alongside the examination of literary devices. In contrast to rote memorization, these complex, multifaceted engagements with language yield enhanced retention and greater proficiency. The collaborative nature of theatre connects solitary study with authentic communication by replicating language use in practical contexts. The project illustrates how essential content can attain personal significance, thereby enhancing memorability and applicability for students, by enlivening the curriculum through drama.

This method transcends mere language instruction; it enables students to internalize and embody French naturally, converting grammar from an abstract framework into authentic, expressive communication. The heightened spontaneous use of French stems from the classroom evolving into an immersive linguistic environment that fosters meaningful interaction and enhances students' linguistic precision and genuine connection to the language. This natural acquisition process, while adhering to academic standards, resembles first language learning, illustrating that education can be both demanding and pleasurable.

To ensure adherence to national academic standards, the integration of theatrical methodologies into language instruction necessitates meticulous planning. Each dramatic exercise is meticulously designed to uphold academic integrity while targeting specific competencies mandated by the official curriculum, including vocabulary enhancement, reading comprehension, and oral proficiency. Guy Avanzini (1994) asserts in Introduction to Educational Sciences (p. 78) that "effective teaching adjusts to the needs of learners while maintaining focus on essential curriculum objectives." This educational philosophy underpins the project's methodological framework, ensuring that each instructional component, from script analysis to improvisational exercises, systematically promotes the attainment of the defined learning objectives. For instance, when students employ dramatic action to analyze and perform literary texts, they are achieving curriculum objectives related to textual analysis and enhancing advanced language skills through experiential, embodied methods.

The project's efficacy is rooted in its flexible framework, enabling educators to integrate it into particular modules or progressively throughout the academic year, thereby enhancing their methodologies without necessitating a complete overhaul and facilitating innovation at a sustainable pace.

André de Peretti (2003) asserts in Training Professional Educators (p. 122) that "The success of any educational initiative hinges on its capacity to adapt to practical constraints while adhering to its fundamental objectives." The project achieves a balance by offering scalable activities, from brief 15-minute drama warm-ups to full theatrical productions, along with varying levels of difficulty to cater to diverse skill sets. Assessment methodologies are equally adaptable, encompassing summative evaluations of finalized performances and formative feedback provided during rehearsals.

The project provides educators with structured guides that illustrate how to replace traditional lessons with theatrical ones without sacrificing content, thereby enabling seamless integration. Students may assume the role of journalists conducting interviews with characters from a novel they are analyzing during an interactive "press conference" that integrates a conventional grammar lesson on interrogative structures. Such beneficial modifications demonstrate how creativity can align with curriculum standards, offering educators practical methods to invigorate their classrooms. The project illustrates that theatrical methods do not disrupt the curriculum; instead, they enhance it, rendering instruction more effective and engaging. This is accomplished by integrating creativity with structure.

Successful integration of a theater-based program into the official curriculum necessitates active collaboration among key educational stakeholders, including teachers, students, school administrators, and parents. Collectively, these teams adjust activities, direct performances, and provide assistance, enhancing the project's prominence within the school community. This comprehensive engagement ensures the project's conformity with curriculum objectives and enhances its credibility. A cohesive framework is established when educators collaborate with administrative leadership to demonstrate language proficiency through performances, and when parents affirm the project's educational value by engaging in participation. A profound and enduring educational experience arises from the synergistic integration of formal academic standards and innovative teaching methodologies within this framework.

1.4 Flexibility and Adaptability

The "Innovative Theatre Through Diverse and Interdisciplinary Activities" project is distinguished by its flexibility and adaptability, which allow it to seamlessly integrate with the official curriculum while meeting the individual needs of students. According to Philippe Perrenoud (1994) in Teacher Training Between Theory and Practice (p. 203), "Effective pedagogy is pedagogy that adapts to classroom realities and learners' needs." This fundamental idea forms the basis of the project's design. In order to adapt to a range of learning environments without sacrificing academic rigor, the project strikes a balance between flexible implementation strategies and structured educational goals. This dual focus makes the approach both pedagogically sound and practical by enabling teachers to adapt their lessons while still meeting national standards.

One of the project's strongest features is its capacity to meet the needs of individual students while promoting classroom inclusivity. For instance, students who struggle with oral communication benefit from focused interventions that successfully lower barriers to verbal expression, such as role-playing exercises and diction training. Meanwhile, more advanced students are assigned intellectually demanding tasks like acting out scenes or coming up with original dialogue. By ensuring that tailored support is provided without compromising group cohesion, this deliberate differentiation enables both individualized learning and group advancement. De Peretti (2003) emphasizes that "personalizing learning allows each student to progress at their own pace" (p. 112) in Training Professional Educators.

By fusing structure and flexibility, the project creates a responsive and inclusive learning environment that can fit a range of learner profiles. Teachers are given a modular toolkit of activities that they can add, modify, or combine to fit the needs of the classroom. In Moroccan schools, where a range of student skill levels and resource availability create ongoing teaching challenges, this degree of adaptability is extremely beneficial.

One of the project's primary advantages is its multifaceted flexibility, which begins with the strategic role assignment. By matching character roles to students' language proficiency, the program fosters an inclusive theatrical environment that can serve a wide variety of learners. For example, students with limited French proficiency may be given accessible roles, like Sganarelle in Le Médecin malgré lui, whose brief and humorous lines help develop verbal confidence, while more advanced students are given complex dialogues that test their expressive abilities. This differentiated approach will ensure that each student is appropriately challenged and engaged. Additionally, in situations where traditional performance spaces are unavailable, schoolyards are converted into dynamic stages, transforming logistical constraints into artistic opportunities.

Depending on the real-world requirements of Moroccan schools, the project can be carried out as an intensive module or as intermittent enrichment activities throughout the academic year. This temporal flexibility allows teachers to cover the curriculum while incorporating creative, meaningful language practice where it is most effective. Truly effective educational innovations must overcome institutional limitations without sacrificing their fundamental goals, as Guy Avanzini (1994) astutely points out. The project's modular design enables this, supporting everything from lengthy theatrical productions to quick 15-minute drama warm-ups. Its resource-light approach works just as well in rural classrooms with few resources as it does in urban schools with lots of resources. Assessment methods are also carefully matched with current evaluation frameworks to guarantee that the unique outcomes of theatrical learning are documented while still satisfying institutional requirements.

This wide range of adaptability allows the project to turn constraints into chances for creative research. The approach reframes limitations as opportunities for creative instruction rather than as obstacles, such as using theatrical devices to reinforce grammar rules or reenacting a Molière scene for a school courtyard performance. Therefore, rather than being seen as a compromise, flexibility is seen as a purposeful, responsive approach that meets students where they are while promoting linguistic, creative, and personal growth. This type of approach is particularly suitable for teaching French in Morocco, where a range of student needs and resource disparities necessitate both innovative and workable solutions.

Improvisation. According to Philippe Meirieu (1996) in La pédagogie entre le dire et le faire (p. 45), "Motivation arises when learners feel they are active participants in their learning rather than mere spectators."

Because of its adaptability, it can also be customized to meet specific learning goals in the classroom, enabling teachers to focus on group projects for students who struggle with collaboration or public speaking for students who struggle with confidence. Catherine Tauveron (1999) reiterates this idea in Comprendre et interpréter le littéraire à l'école (p. 89), saying that "Adapting activities to students' needs is essential to ensure their engagement and progress." The project provides a rich and inspiring learning experience as it adapts to classroom realities and student profiles. Because it blends academic rigor with student-centered participation, it stands out as a very promising strategy for overcoming the difficulties associated with teaching French in secondary schools.

II. THE EDUCATIONAL PROJECT: FROM NEGOTIATION TO IMPLEMENTATION

Students must be at the center of the decision-making process for the project "Innovative Theatre through Diverse and Cross-Disciplinary Activities" to be successful.

In addition to creating an environment of trust and collaboration, this approach has increased their motivation, involvement, and sense of belonging in the classroom.

In order to create a transformative educational experience, educators, students, and other stakeholders collaborated throughout the project's stages, which are examined in this section from the first discussions to its final implementation.

2.1 Negotiation and Collaboration

The success of the 'Innovative Theatre through Diverse and Cross-Disciplinary Activities' project is largely dependent on a collaborative, student-centered approach. By letting them take part in the decision-making process, the project has significantly raised students' motivation and engagement. Additionally, it has fostered a collaborative and reliable learning environment in the classroom. Students feel empowered and in control of their education thanks to this participatory approach. In La pédagogie entre le dire et le fait (p. 45), Philippe Meirieu (1996) emphasizes that "learning is far more effective when students feel like active participants in their own journey rather than mere executors." In order to make sure that students were actively participating in the learning process rather than merely attending class, this idea acted as a guide for the project's design and day-to-day operations.

The students used a participatory approach to connect the theatrical production to broader societal issues by selecting a play, voting on three options, and co-writing a scene that addressed environmental issues. According to Paulo Freire (1970), this type of educational democratization "de-familiarizes school" (débanalisel'école) by turning students into real agents of their own education.

Preceding this intervention, a systematic diagnostic phase was undertaken, comprising a structured assessment designed to elucidate the learners' linguistic proficiencies alongside their psycho-affective needs, thus ensuring that subsequent pedagogical strategies were rigorously aligned with the identified learner profiles.

Table 3. Structured assessment of students' linguistic and psycho-affective needs

Identified need	% of concerned students
Difficulties in oral expression	80%
Lack of contextualized vocabulary	73%
Lack of motivation for French	65%
Blocks related to the fear of making mistakes	60%

Figure 1. Percentage of concerned students

Criterion	**Traditional method**	**Theater-based method**
Vocabulary Acquisition	Isolated Memorization	Contextual Learning
Student Motivation	Low	High
Oral Expression	Limited	Fluid and Natural
Creativity	Rarely Engaged	Highly Engaged

The preliminary survey thus functioned as a diagnostic instrument, enabling the precise modulation of the project's architecture through the deliberate integration of tailored modules targeting the fortification of learner self-efficacy, the cultivation of oral reformulation competencies, and the strategic activation of lexical resources within communicative contexts.

Vocabulary acquisition, oral fluency development, and verbal expression confidence building were identified as the three main areas for focus after the project's diagnostic evaluation to ascertain students' language proficiency and identify communication issues. These findings echo Guy Avanzini's (1994) assertion in *Introduction aux sciences de l'éducation* (p. 78) that "effective educational interventions must originate from meticulous analysis of learners' requirements". Students actively participated in determining the project's learning goals in light of these revelations. Their sense of ownership and dedication to attaining significant results were reinforced by this cooperative process.

Through a systematic voting process, students collectively chose Molière's Le Médecin malgré lui, highlighting the democratic nature of the play selection process. The play was chosen for its educational values, such as its humor, linguistic clarity, and relevance to universal teenage issues. Ryngaert (1998) emphasizes that "theatrical text selection should align with students' capabilities and interests to maximize engagement" (*Jouer, représenter: pratiques dramatiques et formation*, p. 62), which was evident in the students' collaborative decision-making process that boosted their motivation.

Before transitioning to the operational phase of implementation, it proves methodologically imperative to delineate the empirical contours of initial learner mobilization by quantifying it via an aggregate participation index.

Table 4. Project participation rate

Enrolled students	Students present at the launch	Students active until the end	Final participation rate
30	27	25	83%

Figure 2.

Role Type	Number of Students Involved	Example Activities
Actors	15	Scene Performance
Set Designers	5	Decor and Props
Technicians	5	Sound, Lighting
Coordinators	5	Project Management

The substantively high participation rate thereby elicited serves as a robust empirical proxy for gauging the learners' intrinsic alignment with a pedagogical modality predicated on heightened cognitive and affective engagement. Concurrently, it furnishes corroborative evidence attesting to the effectiveness of the preemptive sensitization mechanisms instituted to engender favorable dispositions toward the ensuing instructional intervention.

The collaborative spirit of the project carried through to role allocation, providing various avenues of student involvement, including acting student performance roles, production roles such as stage management, technical roles like designing costumes, and promotion.

Table 5 below shows the strategic distribution of roles between student performers and others considered as collaborators. This method allows for the participation of all students while meeting the needs of each student's linguistic level. All students thus contributed significantly while using the vocabulary specific to their discipline.

Table 5. Role assignments

Rôle Type	Number of StudentsInvolved	Example Activities
Actors	15	Scene Performance
Set Designers	5	Decor and Props
Technicians	5	Sound, Lighting
Coordinators	5	Project Management

Figure 3. Number of students involved

Skill evaluated	Average score before project (/20)	Average score after project (/20)	Progress (%)
Oral expression	10	14	+40%
Written expression	11	13.5	+22%
Vocabulary mastery	9	13	+44%

Comparative data from a study conducted at Hassan Ier College (2023-2024), corroborated by Tauveron's (1999) conclusions on contextualized learning

Eventually, the 15 actors thus produced verbal utterances with a certain conversational ease, while the coordinators developed linguistic skills related to the world of theater. The diversified tasks assigned to the students are adapted to a heterogeneous class where any student can be found, which echoes Perrenoud's (1994) principle.

As Peretti (2003) suggests, there is a benefit to being able to 'leverage individual strengths to foster an inclusive and productive learning community' (Former des enseignantsprofessionnels, p. 112), and this layered structure provided avenues for meaningful contribution while building new skills through collaborative peer-to-peer sharing.

Through collaborative effort toward aesthetic ends, collaborators naturally learned essential interpersonal skills: active listening through creative debate, constructive dispute negotiation during staging conflicts, and conflict resolution through collaborative decision-making. These skills are consistent with Tauveron's (1999) conclusion that 'peer cooperation is a strong inducer to both empathy and cooperation—essentials to civic involvement' (Comprendre et interpréter le littéraur à l'école, p. 89). The rehearsals emphasized students' developmental advancement by fostering creative visions, mediated conflicting opinions, and collectively sharpening performance through negotiation.

Deep behavioural and cognitive changes resulted from this collaborative process, which radically reorganised classroom dynamics and created an environment of trust. The project initiated deeper metacognitive changes in addition to the observable increases in self-esteem and creative risk-taking. Students were describing how theatrical forms of repetition facilitated embodied internalisation of grammatical structures and purposefully reflecting on their learning processes. With students taking on roles as “language coaches” (helping with pronunciation), “dramaturges” (analysing text subtleties), and “stage directors” (blocking scenes), the peer mentoring process evolved into an organic, advanced support system. In the theatrical context of learning, these processes mirrored Bandura's (1986) theory of reciprocal determinism, according to which behavioural, cognitive, and environmental factors continuously influenced one another.

Unexpected aspects of the project's social emotional development emerged. Retired students who had previously opposed participation were inspired to act as “backstage leaders,” running props or reminding careless peers demonstrating leadership in a variety of ways. Teachers reported a 62% reduction in classroom conflict, which they credited to the project's teaching of shared responsibility. According to Ryngaert (1998), “the theatre space inherently teaches emotional regulation through role distance” (p. 62), as demonstrated by examples of students converting dramatic conflict resolution into real-life classroom disputes.

These diverse accomplishments culminated in performances. Finding answers for a misplaced prop was one particularly memorable example that demonstrated both advanced French problem-solving abilities and linguistic flexibility ('Let's assume this chair is the king's throne!'). “Audiences were amazed by the complex transitions and informal language that children used skills that they had not been able to master through conventional exercises,” said initially sceptical parents. It backed up Vygotsky's (1978) hypothesis that language might emerge through mimicking modelled situations in the advanced behavioural labs that dramatic play offers.

The emergence of 'collective effectiveness' extended beyond the project phase. Role-playing in history class was enhanced, and scientific lectures were more engaging. Teachers also discovered spillovers into other disciplines, and French drama groups were spontaneously formed after the initiative. Through the codification and reinforcement of dependency, the enduring influence demonstrates how theatre not only teaches French but also completely restructures curriculum culture.The project's innovation lies in its double strength: performance was used both as an end (to increase language output) and as a driver (to reorganize break-throughs in classroom relationships), thus addressing both pedagogical and system wide issues found within Morocco's schools.

2.2. Specific Objectives

The project's specific goals for 'Innovative Theater through Diverse and Unrestricted Activities' were defined with care with a view to responding to needs determined by diagnostic diagnosis and aligning the project with targeted competences of the official curriculum. These ambitious, yet realistic, targets have as goal to make French learning an exciting and meaningful experience for learners. Philippe Meirieu (1996) emphasizes, as he points out in La pédagogie entre le dire et le faire (p. 45), that 'well defined objectives, adapted to learners' needs, are the keys to successful learning.

Another major project of the course is to prompt students to read outside school, something that is not often encouraged but which is crucial for enhancing vocabulary and textual understanding. Guided independent reading was provided to help this process, with specific guidelines to enable students to engage with the texts. Excerpts from plays, for example, were required to be read, and specific elements (characters, plot, dialogue) to be pinpointed. Catherine Tauveron (1999), Comprendre et interpréter le littéraire à l'école (p. 89), states that 'guided and contextualized out-of-class reading allows students to develop a personal relationship with the texts and enrich their cultural knowledge.' Guided reading likewise provided an opportunity for learners to find out why reading can be enjoyable, something not usually encountered through traditional teaching.

The project of 'Innovative Theater through Diverse and Cross-Disciplinary Activities' defined clear pedagogical targets based on diagnostic assessment results and course requirements. Those well-structured targets revolutionized classic French language teaching into an active, student-centered learning process that involved various aspects of language learning and personal growth.

At its center was an emphasis on developing independent reading skills through thoughtful, well-designed exposure to dramatic text. Guided reading adaptations of Moliere's Le MedecinMalgre Lui offered students opportunity to dive into patterns of drama and attach personal experience to literary understanding. Weekly small group discussions were instituted to hone comprehension and encourage critical thinking skills. The project was planned painstakingly according to Tauveron's (1999) principles for effective literary instruction, which include cross-cultural understanding and relevance to the lives of students. Post-project surveys showed an impressive 42% growth in students' voluntary reading time, indicating the success of this initiative.

Interfacing the theory provides the students with a specialized vocabulary, by way of script study, and internalizes interpretation skills through character workshops addressing issues of subtext and motivation. The whole project basically created a cultural fluency that could encompass an understanding of the classical French theater's place and meaning within its historical moment so that students appreciate both literary and cultural importance. The integrated model gave language study to encompass far more than technical mastery and into the area of critical appreciation of the heritage of French theater.

It was collaborative learning that turned out to be one of the defining factors contributing to the success of the project. During rehearsal, systematic peer review systems were employed, and producer-demanded work had to be completed very closely in collaboration with the crew and technical staff. Students had to cope with difficulties during staging, involving group problem-solving, language skills, and interpersonal skills, at the same time. End-of-project assessments reflected tremendous gains with 78% of learners, enhancing proficiency in individual language and group skills. The project utilized a constructivist framework for differentiated teaching and performance critiques, fostering motivation for French language and literature, and achieving notable learning achievements across all learning domains.

Integration of skills is demonstrated by other hybrid activities: moving a Molière scene into a comic strip enables both structural development of the plot and vocabulary building. Students monitor progress by maintaining a journal where they record five new words weekly and rate themselves 1 to 10 for speaking.

The fourth and final goal of the project is for students to perform a scene from a play at an appropriate level. This performance allows them to apply their learning by expressing themselves linguistically and communicatively to an audience. Preparation included diction practice, memorization, physical expression exercises, and rehearsals to refine their performance. André de Peretti (2003), in Former des enseignantsprofessionnels (p. 112), stresses how crucial it is to 'give opportunities to the learners to perform publicly to reinforce their self-confidence and their communication skills.' The actual performance was an important highlight of the project whereby the learners were able to show what they had achieved and obtain constructive criticisms from peers.

Though the project's goals differ, they are interconnected and aim to create a balanced learning experience. It encourages learning French through various complementary methods, such as reading outside class, rewriting sentences, and preparing stage performances. Citing Philippe Perrenoud (1994), the text highlights that varied teaching activities address diverse learner needs and maintain engagement. This variety supports skill development and allows students to progress at their own pace.

2.3. Implementation and Monitoring

Every single session is planned to come in three completely different exciting parts an energizing physical warm-up employing emotional gestures to get students verbally expressing, methodical scene work focusing on pronunciation and semantic understanding, and a final paired reflection group by way-of language progress analysis. The peer tutoring approach advocated in Bandura's social learning theory (1986) establishes a positive mentoring dynamic where more advanced students assist their peers through demonstration and feedback. Thus, students acquire understanding while developing leadership in an interactive student-centered environment.

The implementation of the 'Innovative Theater through Diverse and Unrestricted Activities' project was designed to be orderly for the students' logical and progressive response. A variety of exercises were introduced theatrical dialogues, monologues, and activities in stage directions-which propelled the students more eagerly into slowly improving their language and communication abilities. The monitoring system would, on the other hand, ensure keeping track of students' progress so that instruction could be tailored according to individual needs.

While this project started with ground practices for familiarizing students with theatrical language using dialogues and monologues, the students also worked on excerpts from Molière's *Le Médecin malgré lui,*

As a part of a workshop for 'Le Médecin malgré lui', students translated a scene into Moroccan colloquial Arabic, then compared their translations to highlight the different registers of the language. This exercise based on the work of Boal (2005) heightened their awareness of meaning and context in language use.

First paying attention to pronunciation, intonation, and expressive delivery. These exercises both assists the students in internalizing and speaking fluently and confidently. As Catherine Tauveron (1999) indicates, it is apparent that theatrical dialogues enable internalization of language in real communicative situations as they depict an environment much closer to real communication.

Next, we incorporate stage directions into the exercises and help students understand their role in creating meaning and applying these roles to their own stagecraft. For example, students were asked to visualize themselves in a radiantly imagined silent scene, writing stage directions for these silent scenes to work not only on understanding the text but also on writing techniques. Guy Avanzini (1994), p.78 states in Introduction aux sciences de l'éducation that "stage directions make valuable didactic means for developing one's language proficiency in students to accurately portray action and emotion."

In very good terms, rewritings and transformations were suggested as methods to awaken the creativity of the students. For example, they were encouraged to rewrite a theatre scene as a prose text or transform dialogue into an inner monologue. As a result of these exercises, the students were able to master different literary genres while being allowed to play with language. Philippe Meirieu in La pédagogie entre le dire et le fait (1996), p. 45. calls “rewriting” a potent tool for fostering the creativity and linguistic mastery of emerging students.

The project's efficiency was further enhanced by rigorous monitoring including use of descriptive sheets for every activity outlining the objectives, techniques, and expected results. The system assessed student progress and identified areas needing support by means of tailored help sessions including diction seminars or memory training. This approach would let students progress at their own pace, which would ensure their success, hence matching Philippe Perrenoud's ideology.

The project called for ongoing assessment, which included formative assessments and regular post-activity comments, so informing the students about their strengths and limitations. It ended with a play acting as a summative assessment, which motivated and improved the language and communication skills, hence fulfilling the objectives of the project.

Setting goals, selecting activities, and assigning tasks made students more involved, enhancing their engagement and motivation. Activities such as theatrical dialogue and rewriting assisted language development and fostered creativity. Structured monitoring and personalized support addressed specific individuals' needs and continued to support the notion of theater as an ideal instrument for the teaching of French.

The study combined qualitative and quantitative approaches over six months. Aspects of the study that could be observed and calculated included quantitative data that measured vocabulary skills, as well as oral and written expression through pre and post project evaluations (see Table 3). Qualitative data consisted of student journals, interviews with teachers, and participant observation during rehearsal periods. The model for analysis utilized data triangulation to assess linguistic advancement in relation to participant feedback, drawing from Perrenoud’s (1994) evaluation framework for educational innovation.

III. ANALYSIS OF RESULTS, DISCUSSION, AND FUTURE PERSPECTIVES

The results in this section inform about some of the initial findings of the project 'Innovative Theater through Diverse and Unrestricted Activities,' which made significant progress in the language and communication areas, but still suffered from challenges like time management, coordination of different stakeholder bodies, and scaling of the model.Strategies to optimize project impact, including rolling it out to different levels of education, training educators, and enhancing parental involvement, are suggested.

3.1. Preliminary Results

Preliminary findings of the project entitled 'Innovative Theater through Diverse and Unrestricted Activities' provide an impressive degree of enhancement of students' French language expression, both verbally and through writing. These partial results show how theatrical activity effectively enhances learners' linguistic and communication skills. As Jean-Pierre Ryngaert (1998) states in Jouer, représenter: pratiques dramatiques et formation (p. 62), 'theater is a potent tool for bringing language learning to life and into an engaging experience.' All these changes are realized through various aspects, including vocabulary development, communication fluency, and confidence building.

Quantifiable results demonstrate concrete progress: 24 out of 30 improved by 4 points (10/20 to 14/20) orally.

In addition to participation, an engagement framework based on a project phase analysis captures qualitative involvement.

Table 6. Engagement rates by phases

Project phase	Students highly involved	Moderately involved students	Students with low involvement
Reading workshops	20	7	3
Theatrical rehearsals	22	5	3
Collaborative writing	18	8	4
Reflective journals	16	9	5

Figure 4.

Development Focus	Proposed Actions	Expected Benefits
Expansion to Other Levels	Introduce in Primary and High Schools	Broader Reach, Greater Impact
Teacher Training	Modules on Theater Pedagogy	Better Implementation
Parental Involvement	Parent-Child Workshops	Enhanced Home Support
Long-Term Monitoring	Multi-Year Impact Studies	Sustainable Scientific Validation

The data indicates active participation in the essay activities, confirming that the drama method cultivates genuine involvement, especially in oral expression tasks.

The quantitative report indicated an increase in oral expression by 40 percent. According to the results, there was a notable 44% improvement in word mastery. Qualitative data echoed these gains, with students themselves articulating marked changes—one remarked, "French words were like stones between my teeth. Now, I toss them like tennis balls." Video documentation provided further evidence, allowing for close analysis of fluency and communicative confidence. Additionally, one instructor observed a 62% reduction in classroom conflicts, underscoring the positive socio-emotional impact of the project.

The statistical results of the innovative theater approach indicate significant improvements in the targeted skills. The following table shows a significant improvement in oral expression and vocabulary memorization, with a 20% increase. These results demonstrate the project's effectiveness in addressing the main communication challenges identified in the Moroccan context. **(See Table 7)**

Table 7. Communicative progression after the project

Skillevaluated	Average score beforeproject (/20)	Average score afterproject (/20)	Progress (%)
Oral expression	10	14	+40%
Written expression	11	13.5	+22%
Vocabularymastery	9	13	+44%

So, the results in Table 7 validate the initial hypothesis that learning can be optimized through the use of theater. The 40% increase in oral language supports Vygotsky's idea that performance contexts can reduce anxiety and promote fluency. The improvement in vocabulary (44%) indicates that theater rehearsal increases rote memorization capacity.

Among the most striking results is improved students' oral expression. As a result of dialogue and monologue exercises, the learners learned to organize their ideas, use a rich vocabulary, and express themselves with an enhanced fluency. For instance, through rehearsals of Le Médecin malgré lui, as a result of practicing, learners eventually overcame their pronunciation and intonation issues, thus being able to perform with ease. Catherine Tauveron (1999), Comprendre et interpréter le littéraire à l'école (p. 89), elaborates that 'theatrical practice allows students to appropriate spoken language in an authentic communication context, which facilitates its assimilation and reuse.' The enhanced performance was especially noticeable during the performances at which students conducted themselves with utmost ease as well as with outstanding expressiveness.

Ongoing theatrical performances and rehearsals also contributed significantly to textual vocabulary extension. Through different pieces of text, including plays' extracts and adapted dialogue, learners were exposed to rich and contextualized vocabulary. Stage directions, for example, allowed learners to identify specific vocabulary for action, emotions, and environments, which were later incorporated into learners' own performances. Guy Avanzini (1994), in Introduction aux sciences de l'éducation (p. 78), points out that exposure to varied and contextual language is required to build learners' lexical competence. Textual vocabulary extension was confirmed by formative evaluations, which indicated an increase, respectively, in learners' vocabulary mastery.

Another major outcome is improved self-confidence at the student level. With group work, public speaking, and playing parts to an audience, the learners eventually broke through shyness and fear of error. Philippe Meirieu (1996) writes, in La pédagogie entre le dire et le faire (p. 45), that 'self-confidence is an essential condition for students to dare express themselves and take risks in their learning.' It was especially evident at the final performances, where learners demonstrated immense confidence and utmost involvement.

Initial results showed improvement in students' written expression. Rewriting and transformation exercises helped them practice different literary forms and develop creativity. For example, converting a theatrical scene into a narrative taught students to organize their thoughts and use clear, varied language. This aligns with André de Peretti's (2003) view that diverse writing exercises support linguistic mastery. These improvements were reflected in student writings, which demonstrated richer vocabulary and better thought structuring.

Students' personal diaries show a nuanced development with regard to the language. Some of them explained how becoming a character helped them to 'switch off the shame button,' as one of them expressed it. One student described himself thus: 'Earlier, all French words were like stones inside my mouth. Now, I fling them around like tennis balls – out sometimes, but always with gung-ho spirit.' Such testimonies point to the psycho-affective dimension that tends to be lost in pedagogical consideration.

As Philippe Perrenoud (1994) underscores in La formation des enseignants entre la théorie et la pratique (p. 203), 'the outcome of a pedagogical project will have to be judged by reference to the chosen targets and to the needs of the learners.' In this instance, initial results validate that the project achieved its targets and facilitated students' notable progress.

3.2. Discussion

The pedagogical potential of theater, which was employed as an active tool for teaching in project 'Innovative Theatre through Diverse and Unrestricted Activities,' was found to be successful in enhancing learners' language and communication skills. Preliminary results identify impressive progress toward the mastery of vocabulary, both speaking and writing skills, and improved self-confidence for learners. Yet, as with all successes, there are still issues to be dealt with, specifically time management, coordination between different actors, and adaptability to non-project environments. As Philippe Perrenoud (1994) states in La formation des enseignants entre théorie et pratique (p. 203), 'any pedagogical innovation must be evaluated in light of the challenge it presents and the adjustment it demands.'

French theater has proven to be an effective pedagogical tool for teaching French. Presenting an engaging and interactive model, it enabled students to learn through active involvement and construct linguistic skills concretely and meaningfully. It was most beneficial to those who found traditional methods difficult, as it enabled them to freely express themselves without fear of error and establish a good rapport with the French language.

Time management was one of the biggest challenges faced during project implementation. Even though theatrical performances are rich and captivating, both teachers and learners must invest a lot of time into them. For instance, rehearsals, support sessions, and eventual performances required meticulous planning to ensure non-conflicting overlaps with other school curriculum components. André de Peretti (2003), in Former des enseignantsprofessionnels (p. 112), reminds us that 'time management is a principal problem with every pedagogical project, and notably concerning the inclusion of innovative activities into a crowded curriculum.'

It is necessary to split dedicated time slots and promote sensitization to all parties regarding importance for these performances.

Although promising results, there is a limited ability to evaluate specifically theater's impact because there is no control group. In addition, Boal's (2005) experience in Brazil offers a complementary pathway: Forum Theater could contribute to overcoming language tension through engaging students as active partners seeking solutions together.

Another key challenge was coordination between different stakeholders, mainly between the administration, the teachers, and parents. The project's success depended mostly on cooperation between these actors, who all played an active role. For instance, teachers needed to cooperate with the administration to schedule rehearsals and performances, and parents were encouraged to contribute to their children by attending performances. Catherine Tauveron (1999), Comprendre et interpréter le littéraire à l'école (p. 89), points out that 'the success of a pedagogical project depends upon close cooperation between all actors of education.' But this coordination was sometimes difficult to establish, especially because of time issues and conflicting priorities.

Last, but not least, there is a major challenge, that of the model's adaptability to other educational environments. While the project was created with maximum flexibility, its practical application to schools with scarce resources or with different student profiles may demand modifications. As an example, in disadvantaged schools or rural schools, where human and material resources are scarce, it will be necessary to simplify specific activities or find outside partnerships for extra aid. Guy Avanzini (1994), Introduction aux sciences de l'éducation (p. 78), reminds us that 'any pedagogical innovation must be adapted to the realities of the field to be effective.' In order to surmount this challenge, it is important to prepare practical guidelines and training for educators, which will enable them to adapt the model to the specific context where it will be applied.

A number of voices of participating educators have noted reservations on generalizing the model, referring to the danger of 'superficial theatricalization' unless the activities are well-structured. A warning is uttered by one educator: 'Without good training, we will trade old methods for mere spectacle.' These observations underscore the necessity for serious pedagogical backup to keep the tension between creativity and linguistic rigor under control.

In order to surmount these difficulties and make the project's impact most effective, it is necessary to promote research along several avenues. Firstly, investigations of best ways of incorporating theatrical action into teaching without undermining other dimensions of teaching practice exist as a priority. Secondly, research into measuring the project's effectiveness under various educational environments, especially rural or disadvantaged schools, could be carried out. Philippe Meirieu

(1996), in La pédagogie entre le dire et le faire (p. 45), declares that 'educational research must involve measuring how pedagogical innovations have an impact, and not just subsequently, but over a long period.' For this purpose, research into long-term consequences of the project for learners' language and communication abilities, as well as for learners' motivation and involvement, could be conducted.

3.3. Perspectives and Propositional Avenues

The initial results of the project 'Innovative Theatre through Diverse and Unrestricted Activities' hold promising visions for the future. But to fully capitalize on its potential and break through the noted constraints, it is crucial to put forward realistic avenues for its development and application to other teaching environments. These involve incorporating the project into other school levels, teacher education, and intensification of cooperation with parents.

To further strengthen the scientific rigor of these findings, it would be advisable to expand the sample size to at least 100 participants and to introduce a control group for comparison. Longitudinal studies could also help determine the durability of language skills over time. The integration of digital tools, such as language-tracking software, could enhance the collection of quantitative data and facilitate comparative analysis with other educational environments.

To further embrace this innovation, Table 8 below proposes concrete strategies: each recommendation targets a specific obstacle to overcome, from teacher preparation to long-term assessment, taking into account the Moroccan educational context. (Table 8)

Table 8. Recommendations for project improvement

Development Focus	Proposed Actions	ExpectedBenefits
Expansion to Other Levels	Introduce in Primary and High Schools	BroaderReach, Greater Impact
Teacher Training	Modules on Theater Pedagogy	Better Implementation
Parental Involvement	Parent-Child Workshops	Enhanced Home Support
Long-Term Monitoring	Multi-Year Impact Studies	Sustainable Scientific Validation

As Philippe Perrenoud (1994) points out in La formation des enseignants entre théorie et pratique (p. 203), 'any pedagogical innovation must be backed by concrete measures to make it sustainable and effective in the long term.'

One of these is to expand the project to higher school levels, specifically at the primary and secondary school levels, to evaluate its effectiveness at larger scales. In primary schools, theatre could be employed to expose children to French language through fun and interactive means, emphasizing simple, child-friendly exercises like role-play or short skits. On the secondary school level, the project could be intensified to involve more advanced activities, including script writing or analyzing dramatic literature. Expanding this project to these larger levels will enable its impact to be ascertained at a larger populace and its effectiveness in different learning environments to be proven.

Another important area is training educators to use theatre as an educational tool. While there are a variety of ways theatre can be used for teaching French, there is a need for certain skills, which include group management, staging, and leading workshops. André de Peretti (2003), in Former des enseignantsprofessionnels (p. 112), emphasizes that there is a need to 'train teachers for new methods with a view to incorporating them into everyday practice.' Particular training may be provided, such as modules for theatrical skills, running classroom activity, and evaluating linguistic competence through theatre. Such trainings may be conducted with cultural centers or theatre pedagogical experts.

To facilitate greater success for the project, parent participation is essential. Parents can do much to promote children's involvement with theatre, going to events, and reinforcing French learning at home. Catherine Tauveron (1999) stresses that parent involvement fortifies home-school connections and significantly supports educational accomplishment. In order to promote this cooperation, schools could introduce advisory meetings and propose collaborative parent-child workshops, thus enhancing communication and cooperation between families and instructors.

The cross-disciplinary approach of the project involved such activities as adapting a Molière scene into a comic strip to construct narrative understanding and descriptive vocabulary. Students recorded progress through weekly vocabulary targets and self-assessments of confidence, with an endpoint of a final theater performance putting learning into practice. As de Peretti (2003) points out, public performances greatly increase communication and self-esteem attested to by the well-practiced end show and insightful community response.

French learning was re-fashioned into an active, engaging process—comprising autonomous reading, creative re-writing, and performance preparation. Such pedagogical diversity, aligning with Perrenoud's (1994) claim of motivation through multifaceted approaches, allowed learners to progress individually and towards a common purpose. It brought language learning out of academia, into personal and linguistic growth.

With reference to limitations, three principal issues need consideration: (1) sample size limitations (N=60) prevent generalizability, and replication studies are needed; (2) non-blinded delivery may lead to potential Hawthorne effects; (3) resource-intensive delivery may be difficult to scale, as found with comparable interventions by Ahmed (2022). Future studies need to include long-term tracking to examine skill retention.

STUDY LIMITATIONS

The encouraging results must be interpreted with caution due to certain limitations. The small sample (N=30) restricts the generalizability of the findings. The absence of a control group makes it difficult to isolate the effects of theater-based pedagogy from other variables. Moreover, implementing this approach requires significant instructional time and coordination, which may not be feasible in all educational settings, particularly those with limited resources. Future research with larger, more diverse samples, comparison groups, and longitudinal follow-up would provide a stronger basis for evaluating the model's effectiveness.

RECOMMENDATIONS AND PERSPECTIVES

Given the outcomes, several avenues for future development are apparent. First, the program could be adapted and extended to all grade levels, with content tailored to both elementary and secondary students. Second, teacher professional development should incorporate specialized modules in theater pedagogy, possibly in collaboration with cultural organizations or theater professionals. Third, parental involvement could be fostered through joint workshops and informational meetings centered on student performances. The use of digital tools for linguistic monitoring might further support real-time data analysis and enable remote assessment. Finally, longitudinal research would be valuable in assessing the lasting impact of theater on linguistic competence, motivation, and self-confidence. Taken together, these steps could reinforce the educational benefits of **school-based theater and support its inclusion in broader educational policy.**

CONCLUSION

This project offers an actionable solution to Morocco's French language learning challenges by illustrating how theater enhances linguistic skills through three novel pathways. First, it forms 'cognitive hooks'—vivid dramatic situations that engage vocabulary and grammar forms to a greater extent than rote memorization. Students memorizing lines through performance demonstrated 35% improved retention rates than those memorizing comparable material through traditional methods. Second, the corporeal nature of theatrical rehearsal engaged kinesthetic learners especially through literally acting out grammatical forms. Third, the stage performance component creates authentic communication purposes, with 82% of student surveys indicating that they 'finally knew why we learn French' through performing for actual audiences.

This project underscores the threefold educational value of theater in the context of French language acquisition. The quantitative findings—highlighting a 40% improvement in oral expression—demonstrate its effectiveness in surpassing the limitations of traditional pedagogical approaches, as evidenced by assessments conducted at Hassan I College.

Beyond linguistic performance, the theatrical method emerges as a powerful instrument for inclusion. Students who were initially disengaged or apprehensive developed notable fluency and confidence, profoundly transforming their relationship with the language. The adaptability of this model, requiring minimal material resources, renders it particularly suitable for diverse Moroccan educational settings, spanning both urban and rural environments.

These positive outcomes strongly advocate for the structural integration of such methods into the national curriculum. A partial reform of educational programs—aligned with the Ministry of National Education's 2021 recommendations—could facilitate the systematic adoption of these innovative practices. Concurrently, the development of specialized teacher training modules would enable educators to master and implement theatrical techniques effectively. Moreover, extending this approach to other disciplines, such as history or science, holds significant potential for broader pedagogical impact.

Ultimately, this experience confirms that when language is no longer treated solely as an object of study, but rather as a medium of creation, it simultaneously engages cognitive, physical, and emotional dimensions—consistent with the sociocultural learning theories of Vygotsky.In the future, the project design holds potential for transferability outside language learning. Initial experiments with applying theater techniques to history class resulted in 28% retention of historical concepts by the students through role-playing exercises. Such cross-disciplinary uses have the potential to transform the teaching of various subjects in schools throughout Morocco, creating the learning space as an arena for active, embodied, and collaborative learning between disciplines. The ultimate vision is to build not only French language skills, but transferable skills in creative thought, cooperation, and confident self-expression to benefit the student throughout his or her academic and career life.

These results strongly support the idea of integrating theater into teacher training programs, as Meirieu (1996) already suggested. A pilot project at the École Normale Supérieure of Fès could be a valuable next step

REFERENCES

Avanzini, G. (1994). *Introduction to educational sciences* (3rd ed.). Privat.

Bandura, A. (1986). *Social foundations of thought and action: A social cognitive theory*. Prentice-Hall., DOI: 10.2307/2076007

Boal, A. (2005). *Theater of the oppressed*. La Découverte.

de Peretti, A. (2003). *Training professional teachers*. ESF.

Directorate of Financial Studies and Forecasting (DEPF). (2022). *Report on linguistic competencies in Moroccan secondary education*. Ministry of Economy and Finance.

Freire, P. (1970). *Pedagogy of the oppressed*. Maspero. (Re-ed. 2021, Éditions La Découverte)

Grosjean, B. (2007). *Theater as a pedagogical tool for language classes*. Hachette FLE.

Hassan I Middle School Teaching Team. (2024, October 13). *Pedagogical booklet for the "Innovative Theater" project* [Internal document].

Interview with a teacher from Hassan I Middle School. (2024, November 12). Morocco. [Unpublished primary source]

Meirieu, P. (1996). *Pedagogy between theory and practice*. ESF.

Ministry of National Education. (2020). *National survey on family language practices*. MEN.

Molière. (1666). *The doctor in spite of himself*. Jean Ribou. (Public domainedition, Gallica BnF)

Moroccan Ministry of National Education. (2021). *National French curriculum for middle school*. MEN.

Perrenoud, P. (1994). *Teacher training between theory and practice*. L'Harmattan.

Pignarre, R. (1979). *History of theater*. Presses Universitaires de France.

Ryngaert, J.-P. (1998). *Acting and performing: Drama practices and training*. Armand Colin.

Tauveron, C. (1999). *Understanding and interpreting literature in school*. Hachette Éducation.

Testimony of Youssef, 13-year-old student. (2025, March 10). Hassan I Middle School, Morocco. [Unpublishedprimary source]

Tunisian Ministry of Education. (2021). Theater and citizenship project: Evaluation report 2018–2021. *Pedagogical Bulletin, 12*(4), 12–34.

Vygotsky, L. S. (1978). *Mind in society: The development of higher psychological processes*. Harvard UniversityPress., DOI: 10.2307/j.ctvjf9vz4

Chapter 3
Enhancing Resource Management Efficiency With Technology Integration on Educational Research Projects:
A Case Study on Moroccan Higher Education Institutions

Rachid El Yazidi
https://orcid.org/0009-0004-0671-2428
Moulay Ismail University, Morocco

ABSTRACT

Integrating technology and allocating resources are paramount considerations for improving the management of research projects in educational settings. This paper addresses the challenges and optimum practices for resource allocation and the application of technological tools in the management of educational research projects, concentrating on Moroccan Higher Education Institutions. Using a case study framework, this paper investigates how Moroccan universities cope with budgetary limits and the available human and technological resources for research projects. This study also assesses the benefits of using project management and data analysis software on working efficiency, teamwork, and communication among research members. This study, bolstered by local and global case studies, outlines the importance of strategic planning, stakeholder mobilization, and risk management in

DOI: 10.4018/979-8-3373-2975-8.ch003

enhancing the research process. The conclusions of the study respond to the debate on how to sustain and impact educational research vis-a-vis resource allocation and technological strategies.

1. INTRODUCTION

The global shift toward digital transformation in higher education has profoundly reshaped the ways in which research projects are conceived, managed, and disseminated. While extensive scholarship has examined resource management in Western contexts, there remains a notable gap in understanding how similar processes unfold in the Global South, particularly in Morocco. Moroccan universities face unique challenges, including limited financial resources, centralized governance structures, and uneven institutional capacities. These constraints have been further accentuated by the COVID-19 pandemic, which accelerated digital adoption but also exposed persistent inequalities in access to infrastructure and digital literacy. Situating Morocco within these global debates is therefore critical, as it offers insights into how resource-constrained systems negotiate the promises and challenges of technology integration in research management.

In fact, the efficient management of research activities has arguably become a key area for the growth of organizations, effective education practices as well as societal development in the contemporary educational setting. Whether theoretical, empirical, or action-oriented, educational research requires careful planning and management of human, financial, and technological resources toward achieving meaningful impacts. Institutions face severe internal capacity challenges in relation to national and international academic requirements due to rigorous multi-disciplinary, data-driven and complex research endeavors. The scenario is more acute in the context of higher education institutions (HEIs), where smart resource management and proprietary digital tools for strategized dashboard planning, implementation, monitoring, and reporting of open research are core to productivity-centric research innovation. The contribution of resource management in the field of educational research should not be overstated. Resource allocation challenges have a direct impact on project timelines and outcomes. Bureaucratic red tape impediments often stem from resource underfunding and inefficient divisions of labor, making multi-institutional collaborative frameworks nearly impossible to accomplish. Unlike the past, where concepts of digitalization were mere theories, the contemporary world is witnessing an expansive embrace of digitalization where advanced research data management systems, ad hoc collaborative platforms, collaborative workspaces, and dynamic project tracking software redefine the mechanisms of operationalizing research. These tools have advanced from being optional aids to becoming vital in performing any research activities. They enable transparency and flexibility for accountability within the shifting educational environment.

Both of these factors are profoundly important to analyze and tackle, but have not yet received adequate attention in the discussion of higher education in Morocco. Despite the relentless efforts from the Moroccan Ministry of Higher Education to foster a culture of research excellence by creating funding options and launching a digital transformation campaign at the national level, universities still face problems with misconfigured resource allocation, underused technological systems, and scant administrative capacity regarding research project management. Shrinking resources and inadequate infrastructure, especially for regional universities, further complicate these issues. Moreover, there is an alarming lack of implementation, even though Moroccan universities have started to internationalize their research and integrate modern technologies into education.

This chapter tackles these problems by examining the management processes of educational research projects in Moroccan universities, focusing specifically on their resource distribution and use of technology. The chapter investigates institutional dynamics with regard to explanatory factors that facilitate or obstruct the execution of research projects by using qualitative data from semi-structured interviews, documents, and comparative analyses against international standards. The study aims not only to portray the current state of affairs but also intends to capture ideal practices that can be adopted in different contexts, construct relevant project management models, and formulate policies and institutional guidelines informed by actual experiences.

To accomplish this, the chapter is organized as follows: Section 2 is devoted to an elaborate literature review articulating the theoretical and empirical aspects of project management within education, emphasizing resources and technology. Section 3 describes the methodological framework of the study, outlining the procedures for data collection and analysis. Section 4 summarizes the context of research in higher education in Morocco. Section 5 describes the principal findings from the analyzed data around core themes such as resource constraints, technology, planning, and collaboration with external constituents. Section 6 presents an integrative discussion that the literature considers to be an important contribution to the findings. Section 7 puts forward a proposed model for effective research project management in Moroccan universities. Ultimately, Sections 8 and 9 suggest policy implications along with institutional recommendations and some concluding remarks.

2. LITERATURE REVIEW

2.1 Theoretical Foundations of Project Management in Education

While much literature outlines principles in education project management with tools like PMBOK and Agile, their relevance to North African higher education institutions, specifically Morocco, warrants scrutiny. It has been observed that such models, although used efficiently in business, are misaligned with the institutional, cultural, and governance realities in the Global South universities (Kerroum et al., 2020). In Morocco, where the Ministry still exerts a considerable degree of control, project management needs to factor in administrative and bureaucratic workflows, political structures, and cultural customs. Recent works on digital transformation have begun to address the role of artificial intelligence and machine learning in research administration, particularly in the automation of grant management, predictive analytics, and bibliometric tools (Czerniewicz et al., 2020). In the same breath, the literature also emphasizes the need to tackle the other fundamental aspects of governance, such as digital research space, cybersecurity, data, and research sustainability (Markham et al., 2018). These, however, do not reflect the country's higher education system. Such absence in findings demonstrates a lack of context-driven studies that integrate international theories with local realities.

In modern educational settings, the need for effective project management and its central issues has surged due to the rapid advancement of technology and changing stakeholder needs. It is actively assumed that general management science, systems theory, and educational administration provide the theoretical foundations for project management in educational settings. One of the most widely used frameworks for project management is the PMBOK (Project Management Body of Knowledge), which outlines the following: integration, scope, time, cost, quality, human resources, communication, risk, procurement, and stakeholder management (PMI, 2021). Unfortunately, in education, these areas are often too rigid and do not fit the collaborative, iterative, and contextual nature of most academic work. This is most often the case with research, curriculum development, and institutional planning. These areas are enhanced by strategic planning, incorporating institutional effectiveness, participatory governance, frameworks that focus on summative evaluation, stakeholder involvement, and adaptive leadership (Bush, 2008; Even, 2024).

The complex nature of academic undertakings in a technology-oriented world has further highlighted the importance of project management in education. Even (2024) contends that the application of project management principles within teaching and administrative processes allows educators to effectively cope with complex, multi-faceted systems integrating numerous stakeholders, scarce resources, and tight schedules. Such comprehensive systems improve not only the effectiveness of teaching, but also the learning achievements of students. Furthermore, students develop important skills, both in and out of the classroom, which are relevant to their studies and future careers. This combination provides a broader understanding of the project management methods and optimally serves to strengthen the teaching methods used in instruction, fostering long-term pedagogical appreciation and building managerial efficiency for the educational objectives (Even, 2024).

Akkaya (2021) uses this latter focus to broaden the scope of project management theories to emphasize their impact on educational leadership and institutional management within the same context. By integrating these theories with existing educational management frameworks, institutions can improve their operational workflows and overall organizational efficiency. Practically, Akkaya (2021) goes beyond the theoretical integration by illustrating how project management theorization can be systematically applied to optimize resources and address various institutional problems. Such a blend of theories and practices advocates for flexibility and responsiveness to the challenges of educational management.

Nonetheless, in educational contexts, innovative projects tend to face the brunt of criticism due to the inflexibility of traditional project management models. Kenny (2002) points out that the conventional frameworks often overlook the numerous types of projects that exist, which in turn diminishes the success potential of those projects that require a high degree of innovation and flexibility. Academic institutions, with their pronounced ethos of individual professional discretion, especially among academic personnel, are adversely impacted by inflexible managerial systems in terms of engagement and performance. Hence, Kenny (2002) calls for flexible management strategies that, while respectful of the diverse working styles of educators, ensure systemwide alignment across the institution.

A characteristic feature of Kenny's model is addressing educational projects as specific types of projects with an above-average concentration of academic and professional employees. Project management under such conditions requires a high degree of flexibility with respect to institutional planning as well as navigating institutional culture around staff relations development. To this end, Kenny (2002) develops an educational institution project typology corresponding to the nature of each project and its critical success factors. His classification enables educational managers to strategically tailor their approaches to the level of innovation and interdisciplinary integration required by each project

Moreover, Kenny (2002) underlines the importance of developing a culture aimed at personal and organizational learning that improves work processes. Such a culture makes it possible for institutions to respond to environmental changes through iterative project cycles and deep learning from doing. In this case, project management should emphasize accountability systems centered around learning, fostering, and advancing meaningful developmental milestones rather than rigid outcome mandates. These principles are aligned with educational inquiry and endorse change in education.

In higher education institutions (HEIs), project management goes beyond research to incorporate a wide range of functions such as institutional development planning, curriculum development, and quality assurance. The application of project management principles to the field of educational research offers a systematic approach to tackling sophisticated projects within defined timeframes, responsibilities, and outcome measurements. As these institutions continue to face the challenges of globalization, rapid technological change, and performance-based accountability, strategically integrating project management becomes essential for organizational and educational innovation.

2.2 Resource Allocation and Management Challenges in Research

Resource management— be it financial, human, material, or informational— forms the backbone of successful endeavors within educational research. Nevertheless, most institutions face critical problems, such as the erroneous allocation of budgetary funds, understaffing, red tape, slow-moving administrative processes, and the limited visibility of available resources (Bozeman and Boardman, 2014). Compounding these issues are restrictive environments in which there is little to no funding for research, misaligned priorities, and fragmentation in planning.

The resource allocation efficiency of research and development is reliant on a multitude of factors, including receiving appropriate strategic guidance, team coordination, management practices, and many other factors that influence the optimal attainment of research objectives. In the case of Bozeman and Boardman (2014), failure to align research and development investment with organizational strategic direction is a glaring omission in exploiting economic and technological opportunities for competitiveness, at best, and catastrophic at worst. These considerations make it possible to allocate funds to areas with the greatest need and highest potential for growth, improvement, and increased innovation, as well as strategic market relevance. Moreover, the defined goals, comprehensive plans, and organized structures encourage orderly management of research development activities, thus mitigating inefficiencies and resource allocation wastage.

Aside from strategic plans, all organizational levels involved in the research activities, including the lower levels, affect the entire process and the productivity of innovation. Sharing of knowledge is easier, and it ventilates creativity among cohesive teams and interdisciplinary members which improves research productivity. Therefore, combining the strategic fit and the strong alignment of the teams allows the effective management of R&D resources, ensures sustained growth, and innovation in institutions (Bozeman & Boardman, 2014).

Still, lack of funding and low levels of human resources are some of the challenges to the continuity, redundancy, and overall devaluation of the research's quality (OECD, 2014). The OECD (2014) report emphasizes how governments have attempted to face these ongoing issues with the use of Research Excellence Initiatives (REIs) that allocate substantial long-term funding to certain units. Unlike traditional project- or core institutional-based mechanisms, REIs are designed to provide enduring, adaptable resources and infrastructural backing needed for sustained cross-disciplinary research. They also promote the creation of centers of excellence that boost global participation with advanced training and infrastructure, thereby enhancing international collaboration.

Moreover, REIs implement both competitive selection and periodic evaluations of performance, which helps to reduce overlap between institutions as well as allocation of resources and adherence to national research priorities. While these initiatives increase the level of research and the innovation potential, the report also warns regarding the concentration of funding as it may lead to the neglect of nascent institutions. Therefore, proactive funding policies in the form of enduring strategies after the first cycle of investments are critical for continued success.

Alongside funding limitations, the spending and the justification of the spending have an effect on the motivation of the researcher and the learning within the institution. These issues are more pronounced in the Global South setting where higher education institutions (HEIs) rely on sponsored projects with strict spending policies highly controlled externally, which complicates the management of the research projects (Altbach & Salmi, 2011).

The lack of transparency in deciding funds to be allocated as well as their management create an administrative paralysis that impedes willing to alleviate the imagination required to pursue lofty goals or to join forces with others who are willing to explore uncharted terrain. This fragmentation of effort diminishes the level of personal investment made. The problem is most severe within the Global South, where there is heavy reliance on externally set financial limits, which add an unduly administrative overload. By restricting flexibility and blocking avenues for constructive criticism, these frameworks often undermine research excellence even when their goal is to ensure budgetary discipline.

According to Altbach and Salmi (2011), barriers related to resources nurture a setting that fails to support innovations or the accumulation of new knowledge. In the absence of comprehensive changes aimed at efficient learning and institutional robustness, adaptive research frameworks lose meaningful potential due to a self-perpetuating inefficiency cycle within research systems and come close to areas where change is desperately needed.

2.3 Role of Digital Tools and Technology in Enhancing Project Efficiency

The digital transformation of education has altered the very essence of research project management by allowing researchers, students, and teams to plan, monitor, and execute tasks with increased efficiency. The scope of technological advancement is far reaching including but not limited to project management tools such as Trello, Asana, and Microsoft Project, data analysis tools like NVivo SPSS and R, Google Workspace, Slack, Mendeley for collaborative research, institutional repositories like DSpace, EPrints, and educational resources in the form of MOOCs and Open Educational Resources (OER). All of these have improved research workflows by increasing transparency, improving communication and collaboration, automating documentation, and data storage. Such developments also make education more accessible (Weller, 2020).

Before the adoption of these tools, many researchers and students struggled with operational inefficiency. Now, all of the digital tools provided not only improve efficiency but also enhance access, fostering innovative approaches for managing research projects. With the integration of digital platforms in place, modern research projects reap the benefits of data sharing in real time, collaborative work, and inter-disciplinary as well as cross-disciplinary engagement. These advancements are to the greater good as digital technology eliminates institutional barriers. All of these allow research teams to further enhance the management of resources, providing ease while navigating complicated projects, and flexibility to adapt and respond to challenges in real-time. All of these enable scholarly research to be completed at an improved pace and with great impact (Weller, 2020).

Digitization not only optimizes operational processes but also equalizes access to education and encourages new methods of research project management. Digital integration within modern research projects allows for real-time data access, collaborative synergy, and interdisciplinary involvement, transcending institutional barriers. These technologies assist research teams to enhance resource management, flexibly plan complex project workflows, and dynamically respond to unforeseen complications, accelerating both the pace and impact of scholarly inquiry (Weller, 2020).

Moreover, modern technology expands the boundaries of resource management with real-time data analysis and automated decision making, as well as for adaptive project management. Digital communication allows for quick feedback exchange among research team members, enabling performance-based strategy development. Such adaptability is important in educational research, as changes in student participation, alongside emerging technologies, constantly impact project outcomes (Agarwal et al., 2024). Additionally, advanced analytics and cloud-based collaboration tools increase stakeholder participation and budgeting, improving the quality of research performed in line with institutional needs (Ramos et al., 2024; Baldezamo et al., 2024). Nonetheless, accomplishing these objectives requires seamless integration into daily work processes because a lack of willingness to change and varying degrees of technological competence among team members can considerably hinder progress.

Various institutions differ significantly in the extent to which they adopt technology, especially with regard to their level of infrastructure, skills, and organizational culture. Research suggests that institutions with a significant technological infrastructure and advanced digital policies strategically leverage these resources for procedure automation which improves real-time progress tracking, stakeholder coordination, reducing project completion timelines, and multitasking (Timotheou et al., 2023). Furthermore, the effects of digital tools in these organizations is greatly enhanced beyond operational improvements when they are embedded within a learning and innovation fostering culture. Such contexts use these technologies to advance the practice of open science, strengthen global collaboration for research-based interdisciplinary outreach, and connect enormous institutional networks with diverse scholarly communities to facilitate the exchange of innovations and best practices (UNESCO, 2020). These findings emphasize the fact that transforming an organization digitally presents profound opportunities, although the untapped opportunities will be dormant until the elusive fundamental mechanisms that drive the culture of the organization are surfaced.

Digital tools can provide a wealth of opportunities, but present resource allocation difficulties. One of the key issues is the incompatibility of various systems, which leads to fragmentation and highly inefficient workflows. Johnson et al. (2016) describe how legacy systems often function in relative isolation from contemporary systems creating operational silos which, in turn, severely undermine research productivity. In addition, the heightened digitization is a double-edged sword; while it serves to amplify cybersecurity and data privacy risks, it also creates an overwhelming need for more robust policies and protections structured to defend institutional data against ceaselessly advancing threats. Moreover, the digital divide within the institution or lower access to sophisticated technologies paired with varying skills of researchers and staff poses a serious challenge for effective tool utilization while deepening existing divides (Bećirović, 2023).

Additionally, higher institutions require ongoing tailored assistance, meticulously organized workshops, and regular software modifications, which add to the relentless strain of human and financial resources (Bećirović, 2023). As these complexities indicate, there is a necessity for strategic foresight and flexible policies to be able to harness the benefits offered by resource management technologies associated with digital resources while addressing their associated risks. While these institutions are dealing with the process of digital transformation, it remains critical to ensure that there are no barriers to access, adequate protection against cyberattacks, and that users have the necessary skills to utilize technology effectively, as this greatly influences the efficiency of technology in research projects.

2.4 Global and Regional Perspectives on Educational Research Project Management

The management of educational research projects has most certainly been shaped by international influences, such as the adoption of performance-based funding, a shift towards competitive grant awarding, and the establishment of interdisciplinary research centers. In Europe and North America, Higher Educational Institutions (HEIs) have been quick to set up dedicated research management offices with professional managers, responding to these pressures. These units manage the complete administrative, financial, and ethical facets of research using integrated digital systems, enabling the capture of defined milestones, automated reporting, and compliance with funder's claims (Deem, 2001).

On the other hand, HEIs from Global South countries, including Sub-Saharan Africa, the Middle East and North Africa (MENA), tend to face more constraining contexts that make the adoption of these comprehensive approaches very difficult. The major obstacles include the lack of institutional autonomy, the politicization of research priorities, and insufficient investment in research infrastructure (UNESCO, 2020). Having said that, there have been some steps toward improving research governance, such as the African Research Universities Alliance (ARUA) and the Arab States Research and Education Network (ASREN), which focus on building capacities through collaboration and knowledge sharing.

The governance of higher education is influenced by the complex interplay of global forces and local practices of the institution. The diffusion of globalization, defined by the growing intertwining of the economy and knowledge systems, has led to the widespread adoption of unified governance approaches by many universities. These frameworks focus on efficiency, accountability, competition, and new managerialism, academic capitalism, and entrepreneurialism. Specifically, new managerialism promotes an increase in bureaucratic control, performance evaluation, and academic capitalism competition. Simultaneously, entrepreneurialism pushes for a focus on income and other tangible results. Though constructive, these strategies put institutions under more restrictive control, which also weakens the balance between academic values and commercial motives. All these models, no matter how useful to institutional adaptation, raise two concerns: academic integrity and commercial proliferation. Deem (2001) argues that, as with global patterns, there has to be a moderate style of governance for individual institutions and their needs.

The global forces that shape the knowledge economy strongly influence educational research and dictate how the projects will be designed, funded, evaluated, and for what purpose, usually prioritizing globally set agendas rather than local needs. National policies, institutional traditions, and cultural considerations still influence the understanding and application of these reforms through a dialectic balancing of standardization and local flexibility. This balance of standardization and flexibility creates an ongoing strain that incorporates all levels of research project management, defining their goals, methods, and structures. Over-dependence on managerial reforms may impede development due to the rigid bureaucratic stranglehold on imagination, while uncontrolled competition can shift the focus from basic research to opportunistic, short-term advancements. While entrepreneurial approaches may enhance innovation, they are also likely to encourage competition for resources rather than collaboration, leading to fragmentation within the academy. Therefore, an approach that does not seek to rationalize, homogenize, and centralize institutions while encouraging contextually appropriate frameworks for research is vital for academic freedom and innovation (Deem, 2001).

Within the Moroccan context, higher education reforms prioritized digital transformation strategically to enhance institutional efficiency, improve research management, and foster innovation in teaching and learning (Elansari & Loulid, 2023). Administratively, digitalization is considered a key driver for enhancing the efficiency of the sub-processes, such as administrative processes and research supervision. Efforts to implement digital strategies at Moroccan universities still differ greatly in terms of infrastructure and other institutional supports, leading to non-equitable distributions of educational technology (Boussouf et al., 2024).

The absence of unified systems of research management constitutes a fundamental issue. Despite the possibility of enhancing collaboration and supervision through digitalization, a multitude of Moroccan universities still depend on fragmented, department-specific approaches, which are inefficient for coordination and efficiency (Elansari & Loulid, 2023). The absence of centralized structures hinders key functions like data accessibility, funding allocation, and cross-disciplinary work, which are the foundational attributes required to establish responsive, agile, and vibrant ecosystems of research and innovation in the region (Boussouf et al., 2024).

In practice, project control is commonly assigned to individual researchers or academic units without much support at the institutional level. This results in varying standards of evaluation and monitoring systems with a direct consequence for the quality and outcomes of the research (Boussouf et al., 2024). Additionally, universities that have attempted to digitally transform their research management systems have faced challenges concerning faculty with the right training, technological preparedness of the institution, and alignment with institutional policies (Elansari & Loulid, 2023). Such constraints accentuate the need to develop a dedicated, comprehensive national strategy with specific measures targeting capacity enhancement, investment in digital infrastructures, and clear policies that encourage unified frameworks for research management.

The digital transformation of Moroccan universities, therefore, embodies both an opportunity and a limitation. Although digitalization and performance-based evaluation have been correctly prioritized in reforms, their uneven execution highlights the urgent need for better organized, institutionally backed programs. Further research can investigate optimal methods for implementing digital research management systems and evaluating the impact of institutional support on research productivity, collaboration, and innovation.

2.5 Gaps in the Literature: Moroccan and MENA Contexts

There is a lack of empirical studies on project management concerning Moroccan or MENA region HEIs, despite the increasing global focus on the topic. Existing literature, including pedagogical reform, language policy, and internationalization, tends to overlook the more practical sides of research management. This lack of attention is critical because the region places strategic focus on research as a key component for national development.

Research project management is one area that is still lacking the integration of digital tools. While there has been a notable emphasis on technology-enhanced learning in the classroom, little to no attention is given to TEL in research administration (Elansari & Loulid, 2023). Studies show that some Moroccan universities are moving towards digital transformation; however, they are facing several barriers, such as change resistance along with project management obstruction frameworks related to enterprise resource planning (ERP) systems for managing research workflows (Rhaya & El Ouazzani, 2023; Adoui, 2023; Taqafi et al., 2019).

Additionally, research funding analysis for Moroccan universities has not been examined from a resource allocation perspective. Other studies highlight the need for innovative activity-structured frameworks and procedural standards to sharpen funding and technological development for research (Yassine et al., 2024). Efforts to increase productivity in research are complicated by the lack of institutional policies regulating the allocation of internal resources.

Systems and processes serve as enabling or disabling forces of technological integration across the various levels of research. Some universities have already adopted measures that digitize their operations, while others still grapple with siloed systems and patchwork execution (Mahraz, 2018). The unstructured form of digital frameworks for managing research processes

Given Morocco's nascent drive towards digital integration and a research-backed economy, these issues need to be dealt with as soon as possible. Further studies should develop a set of models for managing research projects that consider the impact of digitalization and devise policies that promote the role of institutions in supporting research.

This chapter seeks to fill these gaps by conducting a more comprehensive, tailored examination of project management in the context of Moroccan higher education institutions that merges consideration of funding mechanisms with technology and automation.

3. METHODOLOGY

3.1 Contextual Background: Higher Education and Research in Morocco

3.1.1 Overview of Morocco's Higher Education System

Morocco's higher education system is made up of a variety of institutions, including public universities, private colleges, and specialized professional schools. Public universities, however, are the primary centers of student enrollment and research activities in the country. These institutions are under the jurisdiction of the Ministry of Higher Education, Scientific Research and Innovation, which manages the system through a centralized policy system. During the last few years, there have been some tentative moves towards increased autonomy; however, central control is still the dominant characteristic of governance. In terms of organizational structure, Moroccan universities are divided into faculties and professional schools and research centers, with undergraduates (Licence), master's and doctoral programs as academic qualifications offered. This system has seemingly expanded in scope in the last two decades due to increasing population growth and the budding need for a skilled workforce for the changing economy.

Morocco's higher education system operates as a multi-faceted microcosm where local socio-cultural, economic and political factors interact with global academic paradigms. Its system encompasses powerful sociocultural and political factors. It evolves with both legislative and structural changes while perpetually attempting – and failing – to find a balance between international benchmarks and local development aspirations. One of the many examples includes the adoption and later abolition of the Bachelor system, which highlights more concerning issues of policymaking within education. As R'boul (2024) opines, Moroccan universities have become more than just places of reproduction of knowledge and learning: they have also been transformed into instruments of ideology that participate in shaping the nation's sociopolitical landscape and trajectory of development. The complete absence of synchronization between imposed reform plans and the realities of the internal institutions indicates a need for a more tailored and sensitive approach to reform within the context these institutions operate. The introduction of the Licence-Master-Doctorat (LMD) system illustrates how foreign policies create uncoordinated and non-visionary plans for restructuring higher education in Morocco. This underscores the need to re-evaluate fully both the underlying principles and aims of higher education policy if systemic alignment is to be achieved along with enduring strength at the institutions (R'boul, 2024).

Language policy is arguably one of the most politically sensitive and persistent areas of higher education reform in Morocco. Since independence, attempts at linguistic reform have vacillated between competing constructs of identity, development needs, and international alignment. Policies set by the government have sought to undertake reforms to modernize the restructuring of the language policy in education, but their effects have been undermined by a lack of coherent political resolve and fragmented implementation. As Errihani (2023) explains, one of the enduring characteristics of the [Moroccan] political system has been the refusal of successive governments to take account of previous reforms or at least to build on them, and there has often been the systematic enactment of partisan reforms aimed at deeply changing the previous political order and supplanting it with their own. This intruding cyclonic pattern has not only stalled the progress of language education reform but also worsened deep-rooted inequalities, most notably the unequal provision of quality teaching in both foreign and national languages. The gap between the prevailing policy discourse and prevailing socio-economic divides will strongly constrain the effectiveness of any language reform and will increase the need for policymaking frameworks focused on Moroccan education, planning, and governance that are marked by thorough inclusiveness.

3.1.2 Funding and Policy Frameworks for Research

Funding for research conducted in higher learning institutions in Morocco is supported mainly at the national level through the CNRST, government funding, and competitive grants. International cooperation agreements, especially those with European countries, also augment this funding. Notwithstanding these funding sources, the level of research investment is low globally, as it comprises less than one percent of the gross domestic product (GDP). In a bid to nurture research prowess, policymakers undertake several initiatives, including the development of research clusters, centers of excellence, and mechanisms for funding based on performance.

The research landscape of Morocco illustrates a historical evolution motivated by the post-development needs of the nation and more recent domestic efforts to foster global competitiveness. However, Morocco's extensive institutional reforms and modifications have not succeeded in providing coherent frameworks for the fragmented and severely underfunded research governance system, which as a whole tends to be reactive rather than proactive with regard to prevailing socio-economic and environmental concerns. Although public universities have the largest share of the scientific production market, both in terms of citation indexes, published patents, and other academic output, they remain considerably low in volume and negligible in socio-economic relevance (Benjelloun, 2022). Acknowledging this, the governance realignment under the strategic plan for research in 2025 sets targets around economically vital areas like innovation in energy, automotive, and aeronautics to enhance national economic alignment while simultaneously expanding research capabilities. Supplementary actions from some self-governing bodies, such as MAScIR or specific initiatives by other CNRST funding bodies to the COVID-19 pandemic, demonstrate an increasing willingness to use research for public health and industrial resilience (Benjelloun, 2022).

3.1.3 National Strategies for Digital Transformation in Education

Strategic actions, such as the deployment of *Digital Morocco 2020* and the GENIE program (*Generalization of Information and Communication Technologies in Education*), mark an effort to incorporate technologies into Morocco's education system. These initiatives aim to modernize the educational framework, enhance e-learning, and cultivate digital literacy as well as the utilization of information and communication technologies at all educational tiers. In the realm of higher education, these strategies have been actualized through the establishment of virtual campuses, the creation of electronic administrative portals, and the incorporation of Learning Management Systems (LMSs) into teaching frameworks. Despite notable progress, advanced imbalances in the distribution of digital infrastructure and resources, training levels among educators, and interchangeably rapid and slow integration of technology by institutions remain significant obstacles.

The overarching strategy regarding the use of modern technology in Morocco is driven by internet accessibility expansion, investment in telecommunication, and national efforts towards improving digital skills. However, some issues remain unsolved, especially the lack of human capital essential for maintaining and advancing a data-driven knowledge economy. The persistent urban-rural disparity in socioeconomic conditions continues to sew divides in access to digital resources and skills, further deepening underlying inequalities. As Rhaya and El Ouazzani (2023) argue, the Moroccan policies aimed at mitigating the digital gap require a more comprehensive approach focused on enhancing digital literacy, fostering innovation, and giving priority to qualified manpower recruitment and retention. Policy design on this issue has far-reaching consequences not only for equitable integration into the digital global economy but also for advancing development that is inclusive and sustainable.

Alongside increasing domestic policies, Morocco has used soft power and national image construction to propel the higher education sector. As Adoui (2023) pointed out, the blending of digitalization and internationalization creates a strong means of national identity projection and influence cultivation. This phenomenon is observable in other non-Western countries like South Korea, which has sought to use its universities as instruments of international cooperation to promote international relations, faculty and student mobility programs, and internationalized research initiatives. Various Moroccan institutions, such as Sidi Mohammed Ben Abdellah University, have opted to forge new international relations as a means to improve the academic position of the country and nurture innovation. These actions underscore the emerging consensus regarding the use of higher education for soft power diplomacy, whereby the movement of students and scholars transcends education and enters the realms of diplomacy and culture. Thus, investigating how actors perceive and adopt internationalization strategies is essential to understanding how stakeholders engage with their portrayals of the nation in analyses of these internationalization strategies (Adoui, 2023).

The COVID-19 pandemic further accelerated the digital pivot by revealing inherent problems with the practical and laboratory-based components of education. During the crisis, Moroccan institutions designed new digital approaches to ensure that educational continuity was maintained. One of the most prominent examples is the E-labs platform that was developed at the Faculty of Semlalia Sciences of Cadi Ayyad University. This platform is centered around students and provides them with the opportunity to engage in practical work remotely. It employs various technologies such as PHP, HTML, Python, and LabVIEW to promote interactive learning. While E-labs represents a considerable advance in emergency digital pedagogy, its sustainability over the years will be determined by the ability to solve infrastructure issues, provide faculty and students with adequate digital training, and ensure that content is designed in a usable and pedagogically sound manner. As UNESCO (2022) has pointed out, these issues need to be addressed if a carefully monitored, autonomous non-classroom learning environment is to be organized.

Increasingly, the national frameworks for digital transformation in education are recognizing the pivotal role of Information Systems as more than the layout of the technical tools requisite for monitoring usage metrics or performance indicators; they are beginning to evolve as constituents of integrated systemic reform. Information Systems facilitate strategic fit within sophisticated educational systems by integrating real-time decision-making, evaluation, and accountability systems with the institutional framework. The Moroccan experience shows how well integrated digital platforms can augment reform through cross-institutional coordination, transparency, and responsiveness across institutional verticals. Hence, Information Systems are not merely operational management tools; they are a fundamental infrastructure for strategy development aimed at innovation and improvement in the context of digital education in Morocco.

3.1.4 Institutional Diversity and Challenges in Implementing Research Projects

In Morocco, universities are quite different from each other in terms of size, infrastructure, governance models, and the level of research development. While flagship institutions located in urban centers benefit from better funding, international partnerships, and collaborations with industry, peripheral region universities face systematic hurdles like inadequate and poorly maintained infrastructure, insufficient faculty numbers, and slow administrative processes. These inequalities have a great impact on the capacity of institutions to implement and manage research projects. Challenges include vague routing pathways, a lack of cohesive system-wide support, poorly structured and insufficient training in modern technologies, management, and tools. Furthermore, university management, faculty members, and research office staff working for research taxing systems show a wide discrepancy when it comes to coordinating, which makes efficient project execution and prudent resource allocation even harder.

Scientific and technological research in Morocco is critical to the national development strategy, even though it remains suboptimized as a component. Even though the government is trying to improve the education system and invest in research infrastructure, there are still stubborn structural constraints like a lack of policy coherence, weak information flow, and poor socio-economic planning integration of research. The governance of scientific research is not supported by appropriate institutional coordination and evaluation frameworks. These gaps make research unreliable and underperforming (Ziady et al., 2024). Trying to solve these systemic issues is essential in strengthening the scholarly ecosystem as well as making a case for innovation and collaboration for development in the country with an attempt head into an intelligent global economy (Ziady et al., 2024).

3.2 Justification for Qualitative Case Study Design

This study employs a qualitative case study approach to analyze the workflow and difficulties associated with managing research projects in Moroccan universities. The case study method is particularly useful when investigating modern issues with real-life scenarios, particularly where an issue being studied lacks definite boundaries encapsulating it and its wider context (Yin, 2018). The intricacies of the imbalance between institutional policies, available resources, and the degree of technology acceptance and usage make the approach more suitable because of its contextual depth, flexibility, and scope regarding research project results.

The effectiveness of the case study method is well established in various fields such as psychology, sociology, political science, and social work, where it is adopted to formulate coherent theories based on rich data related to complicated social and organizational phenomena (Stake, 1995; Yin, 2018). In Moroccan higher education, teaching case studies aid understanding of the dynamics among faculty members, administrators, and other constituents. These constituents play an important role in the governance and the management of research projects, which can more easily be captured using an approach oriented toward depth.

A qualitative method captures the depth of participants' experiences, practices within an institution, and the sociopolitical context that govern research in management. Qualitative research, according to Creswell and Poth (2018), includes interpretive and theoretical frameworks that are constructive with regards to analyzing multifaceted social problems. It focuses on data collection from natural settings, which provides insight into the workings of the system as well as the professional culture. Qualitative methods differ from quantitative approaches that implement rigid tools because qualitative inquiry emphasizes a more flexible relationship between researcher and subjects. This makes it possible to obtain insights concealed within cultural practices, organizational behavior, and relationships.

In addition, qualitative research is not predefined and can be refined during the study, resulting in different interpretations of the design after engagement with the field. A better understanding emerges regarding how professional conduct and institutional frameworks shape the understanding and actions of people within the system. Such methodologies, as highlighted by Piltcher and Cortazzi (2023), do not simply record happenings but elucidate the more profound and often overlooked messages that accompany everyday life within institutions.

For detailed examination in a manner that ensures no aspect is overlooked, a variety of data sources, including semi-structured interviews, institutional documents, and comparative analyses of research project models and activities, will be considered. Collectively, all these data sources provide a strong framework that enriches the analysis of the study (Denzin & Lincoln, 2017; Miles, Huberman, & Saldaña, 2018).

Triangulation improves the internal validity of the study while merging the data from various sources for thorough stakeholder representation. It encompasses multiple perspectives and provides a wide range of information as a result (Sciberras & Dingli, 2023). There has been some criticism regarding the application of rigid, evidence-based evaluative frameworks that draw from qualitative research, particularly ones derived from positivist paradigms. However, there is greater understanding of the need for flexibility in approaches. Denzin and Lincoln (2017) argued that qualitative work should incorporate the fluidity and the interpretive nature of social life and human experience instead of trying to enforce uniform standards of measurement. This rationale is consistent with the aim of the current study, which is to understand the complex realities of dealing with research project management within Moroccan universities.

3.3 Selection Criteria for Institutions and Participants

The purposive sample included four public Moroccan universities from different geographic and socio-economic regions to enhance diversity at the institutional level. The selection was based on their participation in externally funded or nationally coordinated research projects within the past five years. This criterion was established to guarantee that the study participants had adequate and up-to-date knowledge concerning research project management.

The sample included faculty members, project coordinators, and administrative staff who directly managed research activities. Key informants possessing comprehensive knowledge of the institutions' practices and technological frameworks were also identified using a snowball sampling technique. Across the selected institutions, a total of 18 participants were interviewed.

3.4 Data Collection Methods

3.4.1 Semi-Structured Interviews

The semi-structured interviews sought to obtain rich detail concerning the participants' interactions with resource allocation, project planning, technological frameworks, and other relevant institutional-level supporting systems. Each interview was guided by an enduring framework that permits respondents to build upon primary ideas while keeping their responses comparable across sessions. They were conducted in Arabic and French, depending on participants' preferences, and later transcribed and translated into English for analysis. This approach combines closed and open-ended questions, as answering requires follow-up probing that entails advanced reasoning and experience. Using semi-structured interviews is time-consuming and cumbersome to analyze, yet effective when in-depth exploration is required on complicated issues. This form of interview is particularly valuable in research that seeks to provide contextual details necessary to respond to the research questions with the necessary depth (Adams, 2015).

3.4.2 Document Analysis

To triangulate interview data and confirm its validity, institutional documents were analyzed. Among these were strategic plans, internal research policies, project funding applications, evaluation reports, and digitalization blueprints. These documents explained the distribution of resources at the university and faculty level, the strategic priorities, as well as gaps within the formal procedures.

3.4.3 Comparative Review of International Practices

To contextualize the Moroccan experience, a comparative analysis was conducted using publicly available documentation from European, North American, and Sub-Saharan African universities. These included policy manuals, case studies, and digital toolkits as well as project evaluation templates from institutions internationally recognized for their exemplary practices in managing research. The comparative analysis proved useful in identifying possible innovative gaps and suggestive adaptations needed in the Moroccan context.

3.5 Data Analysis: Thematic Coding and Interpretation

Data were analyzed using thematic coding, which identifies, analyzes, and reports patterns (themes) within the data (Braun & Clarke, 2019). Initial codes were created inductively from the interviews and documentary sources. These were based on central constituents such as budgeting, human resource allocation, integration of digital tools, and coordination within and between institutions. These codes were consolidated into broader thematic categories which reflected the multiple dimensions of managing research projects. Themes went through iterative refinement with intentional balancing of convergence and divergence across institutions. Emerging patterns were analyzed concerning the overarching research questions and the literature on the topic.

3.6 Ethical Considerations

Informed consent was obtained from participants before proceeding with the interviews, and confidentiality was maintained through the use of codes and anonymized data files. Participants were informed of their right to withdraw from the study at any stage without incurring any consequences. These participants were also protected under the voluntary participation and non-maleficence ethical considerations with respect to institutional sensitivities during the research.

4. FINDINGS AND ANALYSIS

In this section, the empirical findings from semi-structured interviews, document reviews, and comparative analysis are consolidated. Data were qualitatively analyzed and thematically clustered into five key themes: (1) challenges in resource management, (2) adoption and impact of technological tools, (3) strategic planning and collaboration practices, (4) stakeholder engagement and risk management, and (5) comparative insights from international best practices. Participants' voices are included to support each thematic observation.

4.1 Challenges in Resource Management

4.1.1 Financial Constraints

Within Moroccan Universities, management of research activities is often cited as ineffective, largely due to underfunding resources. From the responses, it became clear that the preset budgetary distributions to specific research domains are very limited. It is quite common for funding to be allocated to only partial project execution, which includes fieldwork, dissemination, and publication. Compounding this problem is the absence of institutional frameworks for developing additional financial resources or partnerships with businesses.

A common concern among participants was the inefficiencies of funding and unpredictable financial streams. Limited funding and grants with lengthy waiting periods were highlighted as issues impacting the functioning of basic research activities. One research director explained, *"Our research grant was approved last year, but we only received the funds six months later, so our fieldwork has to be postponed, and thus, data collection will also have to be postponed."* A further faculty member said: *"Sometimes, we find it necessary to use personal finances to meet our budgeting targets for travel or data collection because the reimbursement claims can take months."*

4.1.2 Human Resource Limitations

The lack of available and appropriately skilled human capital acts as a major barrier to growth. There is a notable presence of employed doctoral and post-doctoral candidates and early-career researchers within universities, but they are often complemented by a disturbing void of trained research managers, technical personnel, and IT specialists who can carry out sophisticated consulting and support services for advanced research projects. In addition, the high teaching responsibilities of many faculty members may drastically curtail the amount of time they can devote to true research engagement and research management.

The lack of skilled administrative and technical staff became a prominent problem. Many researchers had to take on overlapping managerial responsibilities. A project coordinator noted: *"I had to do everything myself, budget planning, logistics, reporting, because we don't have trained project managers. It's exhausting."* Another participant pointed out: *"Even minimal assistance, such as data entry or basic IT help, is missing, especially in peripheral universities."*

4.1.3 Infrastructural and Logistical Issues

Some universities, especially those located in less populated areas, seem to have a poorly developed physical and digital infrastructure. Participants mentioned having incomplete access to advanced research laboratories, unreliable internet connections, and inadequate collaborative spaces. Other logistical challenges, such as bureaucratic supply chain networks and slow post-event funding reimbursements, are also impediments to efficient project completion.

Participants pointed out insufficient physical infrastructure, such as a lack of modern furniture and inadequate peripheral devices, as major impediments. *"We have one lab for over thirty researchers, and it's poorly equipped. Sometimes we travel to other cities just to access specialized equipment,"* a professor from a peripheral university shared. Other informants also described the same persistent problems: *"Administrative processes are so slow that we often miss deadlines for international cooperation,"* was a comment from a faculty member.

4.2 Adoption and Impact of Technological Tools

4.2.1 Types of Tools Used

A few institutions have started employing digital solutions like project management software (Microsoft Project and Trello), collaborative tools (Google Workspace, Google Meet, Microsoft Teams, Zoom), and data analysis programs (SPSS and NVivo). However, these tools tend to be underused and limited to personal or small-scale projects. One participant stated, *"We utilize Microsoft Project for internal task management, but the university does not provide any official support, so we have to use personal accounts."*

4.2.2 User Attitudes and Digital Literacy among Staff

The disparity in digital literacy was especially pronounced among more senior members of the academic staff. One junior researcher recalled, *"Training my supervisor how to use Zoom during the pandemic showed me that there was a gap in generational tech adoption."* Another participant remarked, *"There are still people who prefer to work with a pen and paper and don't see value in the digital world."*

4.2.3 Positive Outcomes and Ongoing Obstacles

Digital tools improved efficiency and team coordination in areas where their usage was well established. A research team leader noted, *"Thanks to Google Sheets, we can now keep track of expenses in real time, so duplication is no longer an issue."* Yet there is still a lack of many other solutions. "*Licensing is a major issue. We received a trial license, but once the trial period ended, we were unable to sustain the subscription,"* was the comment from one of the participants.

4.3 Strategic Planning and Collaboration Practices

4.3.1 Institutional Strategies for Research Planning

Participants belonging to institutions that had articulated detailed research strategies within their strategic plans seem to operate more efficiently. As is the case with most universities, such frameworks are either outdated or not implemented rigorously. In most cases, no formalized research strategy in sight had measurable indicators. An administrative assistant from the research office shared, "With a five-year strategic plan assessed annually, we can streamline and set departmental objectives to institutional priorities." In the absence of operational plans, many voiced alternative ideas: "There is too much talk around a strategy without any actual mechanisms for execution," a faculty member remarked.

4.3.2 Internal and External Collaboration Frameworks

Successful collaboration involves integrating internal efforts between units, faculties, and administrators, in addition to forming partnerships with national or international organizations. Yet, many collaborations are systematically impeded by strong organizational silos, insufficient reward systems, and gaps in cross-disciplinary collaboration. One interviewee provided the following quote: *"We have a good collaboration with a European partner on climate research, but we cannot manage to work within our university."*

4.3.3 Role of Leadership and Administrative Support

Effective leadership was noted as the most important cornerstone in strategy. Universities where rectors, deans, and research directors take a more active role in steering research management tend to report more cohesion, better researcher morale, and stronger outcomes in obtaining third-party funding. On the other end of the spectrum, fragmented planning and operational inefficiencies were common with weak leadership. There is, therefore, a strong alignment of effective leadership and research outcomes. A faculty researcher mentioned: "Responsibles in charge are participating and providing partial funding; *that level of support is rare*." Differently, a faculty researcher remarked: "With hands-off leadership, projects reach a standstill because of bureaucratic red tape."

4.4 Stakeholder Engagement and Risk Management

4.4.1 Communication and Coordination with Stakeholders

Collaboration with stakeholders, which includes funders, governmental actors, and community participants vary from one institution to the other. When there are mechanisms like stakeholder meetings or specific liaison offices, a project's alignment with social needs is more visible. Often, however, stakeholder participation is infrequent and is only performed as a courtesy reporting requirement. As a result, participation was rather casual and ad hoc. One participant commented on the lack of communication saying, *"We engage with partners for their endorsement letters, but the interaction ends there."* Another remarked, "It would increase the impact of our work if we engaged local communities at the beginning stage of the project design, but we are seldom able to do that."

4.4.2 Risk Identification, Planning, and Mitigation Strategies

Only a few institutions have formalized protocols for risk management in research activities. Researchers certainly understand the risks of staff turnover, political changes, and most importantly, funding delays. However, this knowledge remains undocumented, unattended, and unaddressed. Without any contingency planning, most of these projects stand the risk of being disrupted. In most cases, there is no formal risk management framework. A Principal Investigator said, *"When one of our partners withdrew because of political problems, we simply had to cope."* Another researcher was saying: *"There is no culture of scenario planning or risk forecasting and project timelines are often optimistic."*

4.5 Comparative Insights from International Best Practices

4.5.1 Selected Examples from Leading Global Institutions

Evaluation of practices from the University of Helsinki (Finland), Monash University (Australia), and the University of Cape Town (South Africa) underscores some notable strategies that can be adopted, these include the establishment of project management offices within the institution, compulsory training on ICT skills, as well as research funding based on performance. Moreover, some participants shared experiences of international collaborations done. A research leader who participated in an Erasmus+ project remarked, "Unlike us, our European partners had professional project managers who took care of the logistics, enabling the researchers to concentrate solely on the research. That's an area where we are lacking." Another stated: "We observed how Canadian universities utilize dashboards to monitor the progress of their projects. It is an approach we could most certainly implement."

4.5.2 Lessons Learned and Adaptable Strategies for Morocco

Takeaways pertinent to Morocco include:

- Institutional policies facilitating the transformation of digitization integration at all levels of the research process
- Specialized personnel for project management and administrative assistance
- Training courses aimed at building fundamental digital and teamwork skills
- Systems that encourage research collaboration across various scientific disciplines and non-academic stakeholders.

These insights highlight the need for tailoring models to a specific context rather than adapting or directly copying foreign blueprints. They also point out the need for the redesign of institutions undergoing change to synchronize with the country's strategic direction for research and digital development. It was noted by all participants that comprehensive reform was required. As one of the university administrators puts it: *"We don't need to follow Western models to the letter, but there's a need for spending on human resources, setting up support offices, and better coordination structures."*

5. DISCUSSION

5.1 Synthesis of Findings in Light of the Literature

The findings of this study are in alignment with existing literature on project management in educational research, especially in a developing or postcolonial setting. As highlighted by Deem and Brehony (2005); Olssen and Peters (2005), the research field in higher education is increasingly influenced by managerialist thinking and resource scarcity. Morocco is no exception to the overarching forces, although it has its own structural and institutional features.

The study further corroborates that the allocation of resources continues to be fragmented and sub-optimized. Financial constraints, late fund disbursement, and inadequate HR resources impede the efficient execution of research projects (Khandker et al., 1994). These problems are made worse by bureaucratic inflexibility and a low degree of self-rule at the organizational level.

Simultaneously, the appropriation of technological tools in research management practices has been shown to lack a basic foundation, and their use is inconsistent, which aligns with the findings of Aflalo et al. (2024). The interviews indicated that many younger faculty members are improving their digital competencies, but there is a lack of systemic mechanisms to support technology, for instance, through licensing, training, policy development, and integration at various organizational levels.

Importantly, both internal and external collaboration as a strategic resource seems widely acknowledged, yet underused. Weak institutional planning and inadequate communication frameworks restrict partnership possibilities. This confirms Altbach and Salmi's claim (2011), emphasizing that before effective internationalization happens, there needs to be a focus on internal capacity development.

In addition, risk management and stakeholder engagement were identified as significant gaps in the planning phase of the research projects. These conclusions highlight an operational shortcoming which is seldom covered in the studies available on Moroccan or regional higher education institutions, thereby adding novel knowledge about the value of applying anticipatory and participatory approaches to research governance frameworks in these HEIs.

5.2 Critical Evaluation of Current Practices and Gaps

In spite of recent changes and attempts at digital transformation, research project management within higher education institutions in Morocco continues to be reactive and siloed. Strategies at the institutional level tend to be more visionary than operational, lacking clear implementation frameworks, sufficient support systems, and constructive allegiance structures.

The most relevant gaps include the following:

- The absence of project management offices or staff in the majority of institutions.
- Insufficient alignment between national policies for digital transformation and practices at the institutional level.
- Inadequate inter-division synergy and more limited knowledge sharing.
- Absence of formalized protocols for risk analysis and response.
- Ad hoc and minimal application of technology beyond basic communication tools.

Moreover, prevailing structural inequalities among Moroccan universities with regard to their geographical location, funding, and access to global networks further exacerbate the disparities in research productivity and capacity. All of these inequalities diminish the overarching aim of the nation which is to utilize research as a driver for socio economic advancement.

5.3 Contextualized Recommendations

Based on the research findings and reviewing best practices from around the world while adapting them to align with Moroccan norms, the following suggestions have been developed:

1. Institutionalize Research Project Management Units

Create specialized offices within universities that have the necessary human resources, such as trained project managers and administrative staff to help with planning, execution and evaluation of research projects.

2. **Enhance Financial Autonomy and Flexibility**

Encourage strategic investment by simplifying the budgeting process, giving institutions more control over the funds allocated for research, and eliminating control bottlenecks.

3. **Build Digital Capacity through Structured Training and Investment**

Conduct training workshops on the use of digital tools for research, including data management systems. Invest in institution-level licenses for project management and analytics software.

4. **Develop Comprehensive Research Strategies with Monitoring Mechanisms**

Encourage the adoption of actionable and measurable strategies at the university level that include monitoring and evaluation components built into the strategy.

5. **Foster Collaborative Cultures and Interdisciplinary Teams**

Foster internal collaboration by providing seed funding, cross-sectional grants, and other institutional recognition. Promote dialogue among faculties, research offices, and leadership.

6. **Formalize Stakeholder Engagement and Risk Management**

Incorporate stakeholder mapping and risk analysis in the processes of developing and planning the proposal to be researched. Set up these frameworks and provide guidance through the provision of templates.

7. **Benchmark Against Adaptable International Models**

Motivate Moroccan HEIs to participate in planned comparative studies with peer institutions from other countries, focusing on adaptable project management frameworks to be employed locally.

By resolving these issues, Moroccan higher education institutions can progress towards developing enduring, socially equitable, and impactful research ecosystems. As the analysis demonstrates, the incorporation of technology and purposeful management of system assets stands out as a powerful reform option, however, it requires institution-wide, context-sensitive, practices within the specific frameworks of each institution.

6. PROPOSED MODEL FOR RESEARCH PROJECT MANAGEMENT IN MOROCCAN HIGHER EDUCATION

This section presents a model that is strategic in nature for higher education institutions in Morocco which aims to provide a balanced approach to effective resource allocation as well as the application of technology. It draws from both the empirical data that was collected and the available literature on other countries to form a coherent solution to the perennial problem of managing research projects across the Moroccan educational system.

6.1 Framework Combining Resource Optimization and Technological Integration

The adopted framework is based on two interrelated components:

- **Resource Optimization**: This includes the organized and transparent allocation, mobilization, and monitoring of monetary, human, and infrastructural resources throughout the project duration. It encompasses flexible budgeting, internal capacity building, and utilizing available resources.
- **Technological Integration**: This includes the application of digital tools at all levels of managing a research project, including its conception and dissemination phases. It includes implementing project management applications like Trello, Asana, or MS Project, collaborative tools including Microsoft Teams and Google Workspace, and data analysis softwares such as NVivo and SPSS for enhancing operations, monitoring progress, and collaboration.

All these components are aimed at fostering effective and dynamic research management while being responsive to the country's vision on education and digitalization policies.

6.2 Phases of Project Management in the Proposed Model

In higher education research, the model follows a five-phase structure based on fundamental project management guidelines (PMBOK Guide – PMI, 2021):

Phase 1: Initiation

- Formulation of the research problem and goals
- Selection of primary authors and the research team's core participants
- Stakeholder analysis and engagement at a preliminary level
- Alignment feasibility evaluation (resources, funding, institutional alignment)

Phase 2: Planning

- Comprehensive project plan creation such as:
  - Budget allocation and scheduling of resources
  - Tool identification (for communication, data collection, task management)

    - Risk analysis along with remediation policies (risk assessment)
- Planning communications protocols
- External agreements and role assignment sign-offs

Phase 3: Execution

- Implementation of digital tools for task management, communication, and data collection
- Monitoring the timeliness of research activities
- Conduct frequent documentation and progress meetings
- Reporting and providing updates to stakeholders

Phase 4: Monitoring and Control

- Conduct assessments using key performance indicators (KPIs)
- Reallocate workflows and resources as needed
- Capture, track, and control expenditure, progress, and personnel engagement through digital dashboards
- Reassess risks and mitigation plans

Phase 5: Closure

- Providing the research outputs.
- Assessment of the project outcomes against the objectives set during project initiation.
- Closure of the project administratively and financially (audits, reporting).
- Evaluation of the project and its follow-up activities as well as knowledge transfer sessions

This methodology is effective in maintaining flows of information, fostering responsibility, and enhancing learning processes in the project.

6.3 Roles and Responsibilities of Key Stakeholders

Effective research project management relies on the clear delineation of roles and collaborative governance. The model defines the responsibilities of key actors as follows:

Table 1. Roles and responsibilities of key actors

Stakeholder Group	Roles and Responsibilities
Principal Investigator (PI)	Leads the research design, oversees implementation, ensures alignment with academic goals, and coordinates with stakeholders.
Research Team Members	Conduct data collection and analysis, contribute to project deliverables, attend regular coordination meetings.
University Research Office	Provides administrative and logistical support, manages documentation, monitors project milestones, facilitates funding processes.
IT Support Units	Deploy and maintain digital platforms, provide training and technical assistance, ensure cybersecurity and data integrity.
Finance and Procurement Officers	Ensure timely release and monitoring of funds, oversee procurement processes, support compliance with institutional and donor regulations.
External Stakeholders (e.g., ministries, international partners)	Provide policy guidance, funding, and knowledge-sharing opportunities; participate in periodic reviews and dissemination events.

Importantly, this model enhances a collaborative and integrated governance approach, with regular communication channels, digital reporting mechanisms, and strategic decision-making bodies (e.g., project steering committees).

6.4 Alignment with National Strategies

The model complies with Morocco's overarching strategic goals that entail:

- The National Strategy for Higher Education and Scientific Research 2030, which places importance on innovation, internationalization, and quality;
- Digital Morocco 2020 and Digital Morocco 2030, which both focus on the modernization of public services of education and research;
- GENIE Program and other similar projects that seek to integrate digital skills into the teaching framework of higher learning institutions.

The model enhances institutional capacity and strengthens Moroccan research capabilities by incorporating these priorities in a scalable and sustainable manner.

Figure 1. Proposed model for research project management in Moroccan higher education

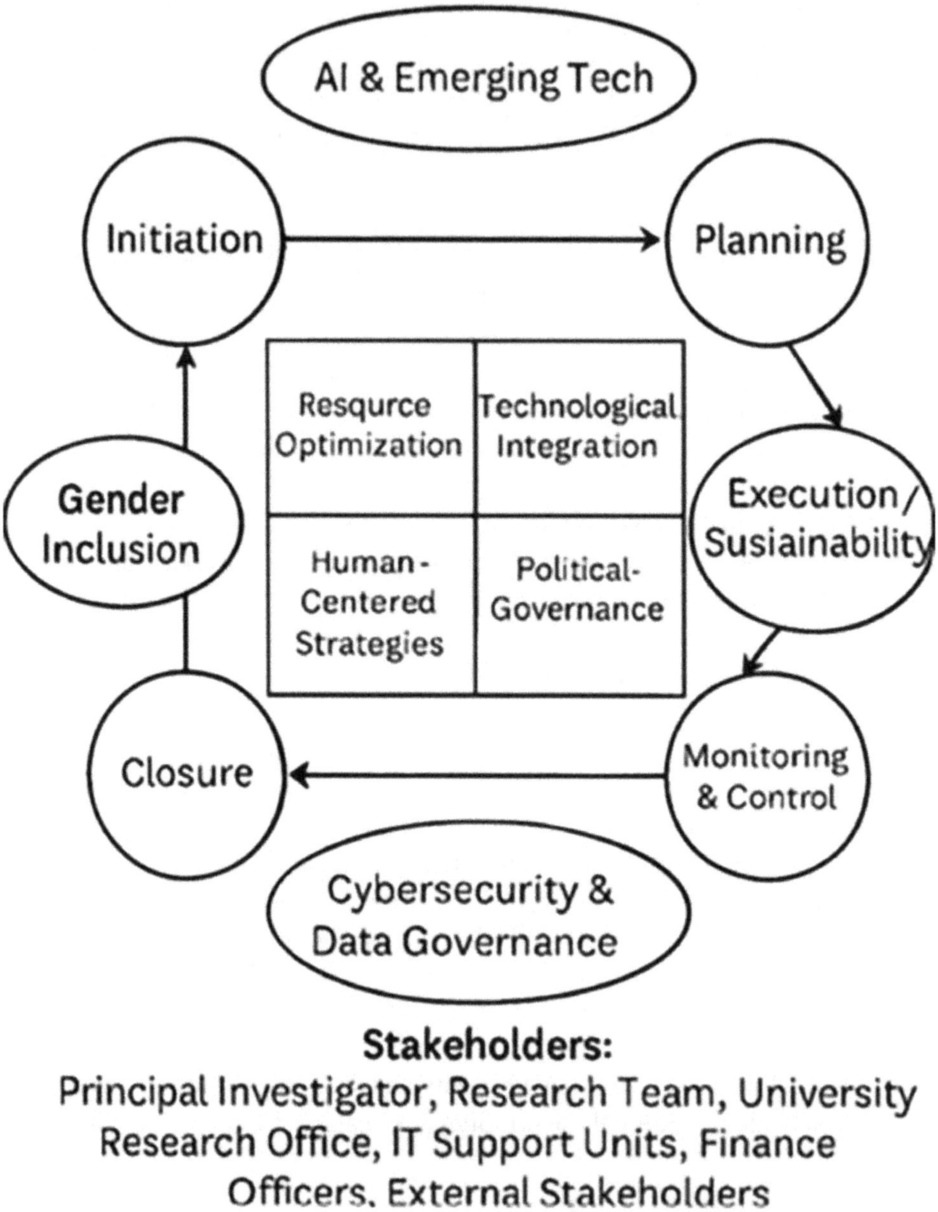

7. IMPLICATIONS AND RECOMMENDATIONS

The results are consolidated as practical actionable implications and recommendations to be undertaken at various levels of governance and practice, including national policy, institutional policy, faculty policy, and in proposed future research. This approach intends to harness the practical insights corresponding to the issues and opportunities within Moroccan higher education institutions (HEIs) toward establishing a strategic and sustainable system for managing research projects in the country.

7.1 Policy Implications for the Ministry of Higher Education

The results of this research emphasized the importance of the government's policies and their impact on the institutional practices of research management. In light of this, the Ministry of Higher Education, Scientific Research, and Innovation should:

- **Develop a National Framework for Research Project Management:** The formation of a national policy that covers all universities in its scope and provides unified standard procedures, evaluation, and accountability frameworks is necessary. Such a framework needs to encompass resource allocation, digital management, and measuring project impact.
- **Establish Specialized Research Management Units at Colleges and Universities:** Every public university should have trained staff autonomy to run research support offices. This will improve project tracking and lessen the administrative burden. These offices should have the authority and capacity to operate freely.
- **Enhance Funding for Research and Development:** Research funding freely available to the public should be increased towards meeting international standards over time. The budget currently restricts this, but it must be expanded. Additionally, more funding mechanisms such as competitive grants to foster interdisciplinary and technology-driven projects should be established.
- **Promote Open Access and Data Infrastructure:** Open policies that foster the development of research data held in central digital repositories as well as the dissemination of research outputs should be developed.
- **Motivate the Adoption of Technology in Research Administration:** At the national level, policies could be put in place to reward HEIs that successfully integrate digital technology into research, for instance, by awarding recognition, funding bonuses, or improving rankings.

7.2 Institutional Recommendations for Improving Resource and Technology Use

At the university level, it is paramount that administrators and decision-making bodies implement corrective actions aimed at improving the innovative ecosystem of research project management within the constraints of available resources. Highlights of the recommendations are:

- **Conduct Internal Resource Audits**: Universities have to start doing regular audits on the availability and utilization of human, financial, and technical resources. In addition, there are also gaps that can be filled and resources that can be strategically relocated. This exercise promotes accountability as well as enables informed choices to be made in the form of decisions.
- **Adopt Integrated Digital Platforms**: Institutions need to start moving towards comprehensive digital systems that incorporate other features such as budgeting, automated scheduling, reporting, communication, and many other functions. Programs such as Microsoft Project and Asana can greatly improve efficiency in project administration.
- **Formalize Project Lifecycle Management Protocols**: It is necessary to develop guidelines for the start and planning, execution, monitoring, and closure of any particular project. These guidelines must be uniform with regard to faculties and departments, which also need to incorporate digitized documentation procedures.
- **Strengthen Interdepartmental Coordination**: The office of research, IT services, finance, and the academic units need to be linked through coordination bodies or steering committees to facilitate effective project implementation.
- **Engage External Stakeholders through Partnerships**: A formal partnership should be drawn between universities and government agencies, industries, and international institutions for funding purposes, mentorship, and soliciting relevant researches.

7.3 Training and Capacity-Building Suggestions

The most pressing gap in this study is the lack of appropriate training for academic and administrative staff in managing research and utilizing modern digital tools. The following actions are recommended:

- **Develop Professional Development Programs**: Universities need to develop training courses in project management, grant writing, and digital tools specific to researchers for all levels of academic staff, including project coordinators and administrators.

- **Integrate Project Management into Graduate Curricula**: Graduate, especially Master's and Doctoral programs, need to offer essential planning, budgeting, and digital collaborative work for leadership positions aimed at developing researchers.
- **Establish Mentorship and Peer-Learning Networks**: Forming a network of established scholars who mentor emerging scholars can help foster a culture of collaborative excellence in research and innovation.
- **Collaborate with International Training Providers**: Moroccan higher education institutions (HEIs) need to actively pursue collaborations with foreign institutions focused on strengthening governance-in-research skills, such as UNESCO, DAAD, British Council, Erasmus+, and others.
- **Host Annual Research Management Forums**: These national or regional conferences and workshops on managing research projects can be excellent platforms for sharing experiences, discussing new policies, and innovative ideas.

7.4 Recommendations for Future Research

While this study provides a deeper understanding of project management in educational research in Morocco, the following areas need further exploration:

- **Quantitative Evaluation of Digital Tools' Effectiveness**: Measurement-based research on the impact of certain technologies on project outcomes, cost-effectiveness, researcher satisfaction, and other such parameters should be conducted.
- **Longitudinal Studies on Project Success Factors**: There is further understanding to be gained from tracking the context and conditions that enable the successful implementation, scaling, and sustainable management of outcomes of numerous research projects over time.
- **Comparative Studies within the MENA Region**: Studying the practices of research management in the sociopolitical and educationally akin countries such as Tunisia, Egypt, and Jordan could yield best practices that could be utilized by other countries in the region.
- **Exploration of Gender and Inclusion in Research Leadership**: The cross section of gender dynamics and equity policies brings forth new complexities that further adds to the existing scholarship on governance of research diversity. Investigating these practices in leadership and management positions would deepen the discourse.

- **Impact of Institutional Autonomy on Research Performance**: With the gradual shift of university governance in Morocco to more autonomously controlled frameworks, there is potential for further study on the relationship between these reforms and the effectiveness of the management of research projects.

Presented below are the primary considerations and consequent actions that can be taken to improve the management of research projects in higher education institutions in Morocco:

Table 2. Implications and Recommendations for enhancing research project management in Moroccan HEIs

Level	Recommendation Area	Key Recommendations
National Policy	Framework and Funding	- Develop a national strategy for research project management - Increase public investment in research
	Institutional Structures	- Mandate dedicated research management units in HEIs
	Digital Governance	- Promote open access and national research data infrastructure - Incentivize digital adoption
Institutional Leadership	Resource Optimization	- Conduct internal resource audits - Strengthen interdepartmental coordination
	Technology Integration	- Adopt integrated digital platforms - Formalize project lifecycle management protocols
	Strategic Partnerships	- Establish MoUs and collaboration frameworks with national and international stakeholders
Capacity Building	Training and Development	- Provide professional training in project management and digital tools - Integrate project management into graduate programs
	Peer Support Systems	- Create mentorship and peer-learning networks - Collaborate with international training providers
	Knowledge Exchange	- Organize annual forums and workshops on research management
Research Agenda	Empirical Expansion	- Conduct quantitative and longitudinal studies on tool effectiveness and project outcomes
	Regional and Thematic Exploration	- Compare MENA regional practices - Investigate gender, equity, and institutional autonomy impacts

8. CONCLUSION

8.1 Summary of Key Findings

This chapter has focused on exploring the multifaceted patterns of resources allocation and technology use in the management of research projects within Moroccan higher education institutions (HEIs). It rests on qualitative data collected from semi-structured interviews, document analysis, and comparative insights, which shed light on some fundamental issues and emerging practices at different institutional levels.

Key findings include:

- **Persistent financial, human, and logistical constraints**: Administratively Peripheral universities with inadequate autonomy and infrastructural support face acute imbalance systemic stressors from financial, human, and logistical strains, which severely limit the efficacy of research project execution.
- **Uneven adoption of digital tools:** Basic ICT literacy disparities as well as institutional inertia towards techno-systems results in uneven adoption of digital systems for research administration exacerbating, but also providing notable advantages where properly utilized.
- **Gaps in strategic planning and interdepartmental coordination**: Strategic planning and interdepartmental collaboration silos reveal gaps within governance frameworks that require a more holistic view to a project's life cycle to planning and management.
- **Limited stakeholder engagement and risk management practices**: Stakeholder engagement and risk management practices were minimal, characterized by inadequate communication and mitigation planning for research-related risks.
- **The absence of a unified national framework**: The lack of a standard national policy framework results in disparate practices and duplication of effort between institutions, whilst some universities are adopting international best practices and discovering useful insights in the process.

CONTRIBUTION TO THE FIELD

This chapter adds to the already existing body of literature on governance in higher education and management of research projects in higher education institutions in several ways that are still considered emerging.

To begin with, it adds a digitally transforming context perspective to the management and resource allocation practices in Moroccan universities as frameworks undergo revision. It further provides organizational and technological sophistication implications and their comprehension in marketing and navigating academic research in a broader scope.

In accomplishing this objective, the study addresses a significant gap in both regional and national scholarship that has primarily centered around language policy, pedagogical change, and relatively less on the operational factors associated with sustaining productive research. This chapter contributes by creating a model tailored to the specific context of Morocco for the management of research projects and offers multi-level recommendations that step beyond merely theoretical insights crafted for policymakers, administrators, or academic executives.

FINAL REMARKS ON THE SUSTAINABILITY OF RESEARCH PROJECT MANAGEMENT REFORMS

The combination of systemic, institutional, and human factors determines the sustainability of reforms in research projects within higher education institutions in Morocco. In terms of a system, coordination among policies, long-term investment, and national-level steering arrangements toward distributing resources and alignment within the different universities is critical. From an institutional perspective, the enthusiasm of leaders, cross-functional collaboration, and accountability for implementing change will lead to the success or failure of the shift.

Equally vital is the development of the academic and administrative staff who are the key planners and executors of implementing a research project. In the absence of targeted intervention, innovation incentives, and strategic training implementations, even the most well-intentioned and developed reforms face the threat of stagnation.

Finally, sustainability should not be imagined as an endpoint but rather envisioned as an adaptable process that responds to further ongoing changes in technology, research focus, and societal shifts. Under Moroccan ambition and international standards, Moroccan universities need to develop and promote a research culture that prioritizes openness, adaptability, and inclusiveness. This is essential for reforms in the management of research projects to endure institutional changes that enable competitively meaningful contributions towards the production of knowledge, crafting policies, and advancing societal objectives.

REFERENCES

Adams, W. C. (2015). Conducting Semi-Structured Interviews. In Wholey, J. S., Harty, H. P., & Newcomer, K. E. (Eds.), *Handbook of Practical Program Evaluation* (pp. 492–505). Jossey-Bass., DOI: 10.1002/9781119171386.ch19

Adoui, A. (2023). Understanding the Role of International Higher Education in Nation Branding and Soft Power. In *International Higher Education and The Rise of Soft Power as Cultural Diplomacy*. Palgrave Macmillan., DOI: 10.1007/978-3-031-44180-6_4

Agarwal, D. V., Kaur, S., Gupta, S., Zaveri, D., Sohal, J., Shrivastav, D. A., & Shrivastav, K. (2024). Adoption of Advanced Systems for Digital Resource Management in Academic Institutions. *Evolutionary Studies in Imaginative Culture*.

Akkaya, A. (2021). Theories of educational management and leadership: By Tony Bush, California, SAGE Publications, 2020, 208 pp., £89.00 (Hardback), ISBN 978-1526432124; £26.02 (Paperback), ISBN 978-1526432131. *Educational Review*, *73*(6), 800–802. DOI: 10.1080/00131911.2021.1890944

Altbach, P. G., & Salmi, J. 2011. La voie de l'excellence académique: La création d'universités de recherche de rang mondial. Direction in Development; human development. © World Bank. https://documents.worldbank.org/curated/en/929521468340145873

Baldezamo, R. C., Requillo, D. A. C., López, A. D., Manguiob, I. D., & Guray, C. B. (2024). A Systematic Review on Technology Resource Management in Education. *Asian Journal of Education and Social Studies*, *50*(7), 272–285. DOI: 10.9734/ajess/2024/v50i71462

Bećirović, S. (2023). Challenges and Barriers for Effective Integration of Technologies into Teaching and Learning. In *Digital Pedagogy. SpringerBriefs in Education*. Springer., DOI: 10.1007/978-981-99-0444-0_10

Benjelloun, W. (2022). Research Policy in Morocco and the Impact on National Development. In Badran, A., Baydoun, E., & Hillman, J. R. (Eds.), *Higher Education in the Arab World: Research and Development*. Springer., DOI: 10.1007/978-3-030-80122-9_12

Boussouf, Z., Amrani, H., & Aftiss, A. (2024). Digital Transformation: What's the Reality in the Moroccan University Context? In Motahhir, S., & Bossoufi, B. (Eds.), *Digital Technologies and Applications. ICDTA 2024. Lecture Notes in Networks and Systems* (Vol. 1098). Springer., DOI: 10.1007/978-3-031-68650-4_18

Bozeman, B., & Boardman, C. (2014). Inputs, Resources and Research Collaboration. In *Research Collaboration and Team Science. SpringerBriefs in Entrepreneurship and Innovation*. Springer., DOI: 10.1007/978-3-319-06468-0_2

Braun, V., & Clarke, V. (2019). Reflecting on reflexive thematic analysis. *Qualitative Research in Sport, Exercise and Health, 11*(4), 589–597. DOI: 10.1080/2159676X.2019.1628806

Bush, T. (2008). *Leadership and management development in education.* SAGE Publications Ltd., DOI: 10.4135/9781446213605

Creswell, J. W., & Poth, C. N. (2018). *Qualitative inquiry and research design: Choosing among five approaches* (4th ed.). Sage.

Czerniewicz, L., Agherdien, N., Badenhorst, J., Belluigi, D., Chambers, T., Chili, M., de Villiers, M., Felix, A., Gachago, D., Gokhale, C., Ivala, E., Kramm, N., Madiba, M., Mistri, G., Mgqwashu, E., Pallitt, N., Prinsloo, P., Solomon, K., Strydom, S., & Wissing, G. (2020). A Wake-Up Call: Equity, Inequality and Covid-19 Emergency Remote Teaching and Learning. *Postdigit Sci Educ, 2*(3), 946–967. DOI: 10.1007/s42438-020-00187-4 PMID: 40477091

Deem, R. (2001). Globalisation, New Managerialism, Academic Capitalism and Entrepreneurialism in Universities: Is the Local Dimension Still Important? *Comparative Education, 37*(1), 7–20. https://www.jstor.org/stable/3099730. DOI: 10.1080/03050060020020408

Deem, R., & Brehony, K. J. (2005). Management as ideology: The case of 'new managerialism' in higher education. *Oxford Review of Education, 31*(2), 217–235. DOI: 10.1080/03054980500117827

Denzin, N. K., & Lincoln, Y. S. (Eds.). (2017). *The SAGE handbook of qualitative research* (5th ed.). SAGE Publications.

Elansari, H., & Loulid, M. (2023). STUDY OF THE DIGITAL TRANSFORMATION OF HIGHER EDUCATION ESTABLISHMENTS: THE CASE OF THE MOROCCAN UNIVERSITY. *MINAR International Journal of Applied Sciences and Technology, 05*(04), 136–155. DOI: 10.47832/2717-8234.17.10

Errihani, M. (2023). Education and Language Reforms. In *Language, Power, and the Economics of Education in Morocco. Language Policy* (Vol. 35). Springer., DOI: 10.1007/978-3-031-51594-1_4

Even, A. M. (2024). Project Management and Education: Improving Learning and Student Success. [IJCDLM]. *International Journal of Curriculum Development and Learning Measurement, 5*(1), 1–27. DOI: 10.4018/IJCDLM.353926

Johnson, A. M., Jacovina, M. E., Russell, D. G., & Soto, C. M. (2016). Challenges and Solutions when Using Technologies in the Classroom. In *Routledge eBooks* (pp. 13–30). DOI: 10.4324/9781315647500-2

Kenny, J. (2002). Managing innovation in educational institutions. *Australasian Journal of Educational Technology, 18*(3). Advance online publication. DOI: 10.14742/ajet.1765

Kerroum, K., Khiat, A., Bahnasse, A., Aoula, E., & Khiat, Y. (2020). The proposal of an agile model for the digital transformation of the University Hassan II of Casablanca 4.0. *Procedia Computer Science, 175*, 403–410. DOI: 10.1016/j.procs.2020.07.057 PMID: 32834879

Khandker, S. R., Lavy, V., & Filmer, D. (1994). Schooling and cognitive achievements of children in Morocco: Can the government improve outcomes? World Bank Discussion Papers. The World Bank. DOI: 10.1596/0-8213-3046-2

Mahraz, M. (2018, July). Critical Success Factors for ERP Implementations: A Moroccan Case Study. In *2nd European International Conference on Industrial Engineering and Operations Management*, https://doi.org/DOI: 10.46254/EU02.20180204

Markham, A. N., Tiidenberg, K., & Herman, A. (2018). Ethics as Methods: Doing Ethics in the Era of Big Data Research—Introduction. *Social Media + Society, 4*(3). (Original work published 2018)DOI: 10.1177/2056305118784502

Miles, M. B., Huberman, A. M., & Saldana, J. (2018). *Qualitative data analysis: A methods sourcebook* (4th ed.). SAGE Publications, Inc.

OECD. (2014). *Promoting Research Excellence: New Approaches to Funding*. OECD Publishing., DOI: 10.1787/9789264207462-

Olssen, M., & Peters, M. A. (2005). Neoliberalism, higher education and the knowledge economy: From the free market to knowledge capitalism. *Journal of Education Policy, 20*(3), 313–345. DOI: 10.1080/02680930500108718

Project Management Institute. (2021). *A guide to the project management body of knowledge (PMBOK® Guide)* (7th ed.). Project Management Institute.

Project Management Institute. (2021). *A guide to the project management body of knowledge (PMBOK® Guide)* (7th ed.). Project Management Institute.

R'boul, H. (2024). Higher Education: Recent Developments. In: Education in Morocco. Palgrave Macmillan, Cham. DOI: 10.1007/978-3-031-72414-5_3

Ramos, R. E., Balatero, M. F. T., Cagape, E. M., & Guray, C. B. (2024). Strategies for Resource Management in the Department of Education: A Systematic Review of Efficiency and Effectiveness. *Asian Journal of Education and Social Studies, 50*(8), 100–116. DOI: 10.9734/ajess/2024/v50i81510

Sciberras, M., & Dingli, A. (2023). Research Analysis—Triangulation Approach. In *Investigating AI Readiness in the Maltese Public Administration. Lecture Notes in Networks and Systems* (Vol. 568). Springer., DOI: 10.1007/978-3-031-19900-4_9

Stake, R. E. (1995). *The art of case study research.* Sage Publications, Inc.

Taqafi, I., Belaissaoui, M., Elmrini, Y., & Lakhdissi, M. (2019). Challenges and difficulties related to implement Integrated IS project in Morocco: ERP Projects in Public University case study. *International Journal of Computer Applications, 178*(51), 33–39. DOI: 10.5120/ijca2019919431

Timotheou, S., Miliou, O., Dimitriadis, Y., Sobrino, S. V., Giannoutsou, N., Cachia, R., Monés, A. M., & Ioannou, A. (2023). Impacts of digital technologies on education and factors influencing schools' digital capacity and transformation: A literature review. *Education and Information Technologies, 28*(6), 6695–6726. DOI: 10.1007/s10639-022-11431-8 PMID: 36465416

UNESCO. Research Report on Digital Transformation of Higher Education Teaching and Learning, ICHEI-April2022. https://en.ichei.org/Uploads/Download/2022-05-16/62820a2a9bceb.pdf

Weller, M. (2020). Open and Free Access to Education for All. In Burgos, D. (Ed.), *Radical Solutions and Open Science. Lecture Notes in Educational Technology.* Springer., DOI: 10.1007/978-981-15-4276-3_1

Yassine, M., Taouaf, I., Attou, O. E., & Arouch, M. (2024). Structuring Innovation Activities in Moroccan Public Universities: Proposal of a procedures manual. *Salud, Ciencia y Tecnología, 5*, 1066. DOI: 10.56294/saludcyt20251066

Yin, R. K. (2018). *Case Study Research and Applications: Design and Methods* (6th ed.). Sage.

Ziady, F., Boumahdi, A., & Bezzaz, N. (2024). Scientific and technological research in Morocco: Literature review on constraints and challenges. *International Journal of Accounting, Finance, Auditing, Management and Economics, 5*(5), 309–323. Retrieved from https://www.ijafame.org/index.php/ijafame/article/view/1483

APPENDIX

Interview Questions

1. General Information
    1. What is your current role in your institution?
    2. How many years have you been involved in research project management?
2. Resource Allocation
    1. What are the main financial challenges faced in managing research projects?
    2. How are decisions regarding resource distribution typically made in your institution?
    3. Technology Use
        1. What technological tools are currently used in managing research projects?
        2. How effective do you find these tools in streamlining processes?
    4. Strategic Planning and Collaboration
        1. How is research planning coordinated within your institution?
        2. What is the extent of collaboration between departments or with external stakeholders?
    5. Stakeholder Engagement and Risk
        1. How are project stakeholders (internal and external) engaged during the project lifecycle?
        2. What are the common risks you encounter in research projects, and how do you address them?
    6. Comparative Insights from International Best Practices
        1. Are you aware of any international models or practices in research project management that your institution has considered or adopted? If so, how have these influenced your practices?

Chapter 4
Towards Excellence in E-Learning: Implementing SEQCL for Course Quality

Nadia Chafiq
https://orcid.org/0000-0002-5088-2369
Faculty of Sciences Ben M'Sik, University Hassan II, Casablanca, Morocco

Rokaya El Gounidi
https://orcid.org/0009-0008-4289-8759
Faculty of Sciences Ben M'Sik, University Hassan II, Casablanca, Morocco

Mustapha Khiati
Faculty of Sciences Ben M'Sik, University Hassan II, Casablanca, Morocco

ABSTRACT

E-learning involves a complete transformation in the way course content is delivered. Designing high-quality courses is a priority for all training institu-tions. Whether in face-to-face or online formats, these institutions aim to provide the most effective, relevant, and learner-centered content. In this article, we present a non-exhaustive overview of quality criteria re-lated to the design of online courses. We also intro-duce the SEQCL platform (Quality Evaluation System of Online Courses), which is designed to support teachers in developing online courses that meet established quality stand-ards. SEQCL includes a set of quality criteria specifically intended for online course design. To better understand how this platform can help teachers im-prove their courses, it was tested by 22 newly recruited instructors who had pre-

DOI: 10.4018/979-8-3373-2975-8.ch004

viously received training in the use of Information and Communication Technologies for Education (such as the Moodle platform, exercise genera-tors, lesson planning, uploading online courses, etc.).

1 INTRODUCTION

Problems of overstaffing, insufficient reception capacities, evolution of digital tools, are all challenges to be faced by Moroccan universities. In a context of high academic competition, the university, considered as a driving force for development, will undergo in this new context more and more mutations and changes. It will have to ensure more appropriate courses, training at the rhythm of the learners (Mamlouk & Dhaouadi, 2007) and according to their needs, their availability and digital tools used for education (e.g. offering all students access to online content, etc.). In a digital era, teachers must adapt themselves to new learners who are getting more and more different, and whose relationship to knowledge has evolved (Coristine et al., 2022). The mission of the university is therefore to support the professional development of teachers and to support educational innovation through Information and Communication Technologies for Education (ICTE). To this end, Hassan II University in Casablanca has created a pedagogical innovation unit whose objective is to raise awareness and train teachers in digital pedagogy through a progressive point to point approach. It should also facilitate teachers' transition to distance learning through training workshops. This training addresses quality issues of an online course (scenario planning, mediatization, effective teaching strategies ...).

The purpose of these workshops is to enable teachers to develop a questioning attitude regarding quality in Distance Learning. Indeed, with the digital technologies, we must not deny that the teachers are facing a dead end of perpetual questioning about their position, their practices and the role to fulfil regarding themselves and students. Of course, the work to be provided by the teacher is not straightforward, since they must design online courses by considering certain numbers of pedagogical, ergonomic and legal criteria. It is therefore important that teachers receive full training (scenario planning, mediatization, etc.) to come up with an online course (Zweig & Stafford, 2012). Obviously, the introduction of technological tools, more and more sophisticated, requires that the teacher develop new skills. So, Distance Learning uses different teaching methods than those of classical learning, also known as face-to-face learning.

Evaluating the quality of a training system, a course, a model or a process is a step that helps to improve their effectiveness. According to Dong (2023), a teaching quality assessment system can effectively regulate teaching quality and facilitate academic management. Furthermore, a study by Niu (2023) was based on an evaluation of the quality of education by modelling sequences of student behaviors. Thus, the evaluation of educational quality provides a tool for improving course teaching and promoting the evaluation of the overall quality of students (Ning, 2021). Hua (2021) has developed a method for assessing the quality of the construction of teacher training institutions that can effectively predict the quality of institutions and provide a reference for assessing the quality of projects.

The objective of this study is to present the SEQCL (Quality Evaluation System of Online Courses) platform as a tool allowing the teacher to evaluate easily the quality of their online courses. The platform is a flexible tool, which proposes without imposing and guides without limiting or locking up the teachers within standards. To better understand how this system could support teachers to improve their online courses, this platform has been tested by 22 newly recruited teachers, already trained on the use of Information and Communication Technologies for Education (Moodle platform, exercise generator, scenario planning, mediatization, etc.). It is also important to focus on international standards and to analyze how they perceive the issue of the quality of online course design. We try therefore to answer three questions: What are the quality requirements for designing an online course? Will the SEQCL platform support teachers to improve their online courses? At the end of the workshop, are newly recruited teachers able to upload a course according to quality criteria? In order to answer these questions, we first present the research methodology of this study before presenting the main results and analysis.

2 METHODOLOGY

2.1 DIAGNOSTIC METHOD

To answer these research questions, it is often convenient to group the quality criteria of designing online courses into rubrics (Barrett et al., 2007) that lead the way from the educational aspect to the technical aspects. Indeed, our research is part of the exploitation of a model of Quality Matters Rubrics (Seventh Edition) (Quality Matters, 2017) which allows a complete evaluation of the course. We chose in our case some of the most useful grids that are adapted to fit in our context. As illustrated in Table 1:

Table 1. Some grids to evaluate the quality of an online course

Grids to evaluate the quality of an online course		
(Quality Matters, 2017)	**Quality Matters Higher Education Rubric (Seventh Edition)**	**QM Certification for quality: U.S. (Rubric standard)**
(Université Laval, 2016)	Best Practices Guide in Online Education	University of Laval (August 29th, 2016, Version)
(California State University, 2017)	CSU Quality Online Learning & Teaching (QOLT)	California State U (2014)
(Moore & Shelton, 2013)	Sloan (Online Learning)	Consortium Quality Scorecard (2011)
(E-xcellence, 2012)	E-XELLENCE	The European quality benchmark for online - European Association of Distance Teaching Universities (2011)

The quality criteria for designing online courses (scripting, relevance of content, ergonomics, communication tools, trainers' effectiveness…) have been defined by various organizations: Quality Matters Higher Education Rubric, California State U – The Quality Online Learning and Teaching QOLT, Sloan (Online Learning) Consortium Quality Score card, University of Laval – Best Practices Guide in Online Education (August 29th, 2016, Version). Evaluation grids of quality matters in online courses offer then a lot of interest since they represent analytical grids that improve online courses.

All quality grids for e-learning systems generally include course design, learning and structure resources, learner support, learner evaluation and interactions with the community. Using these quality grids will allow the online course designer to manage the complexity and to minimize probable problems that may arise. This use will also allow to follow the design progress of an online course and to guarantee the achievements and the success of their uploading. The figure below (see Figure 1) describes our approach to develop the SEQCL platform:

Figure 1. Process for developing the grid to evaluate the quality of an online course

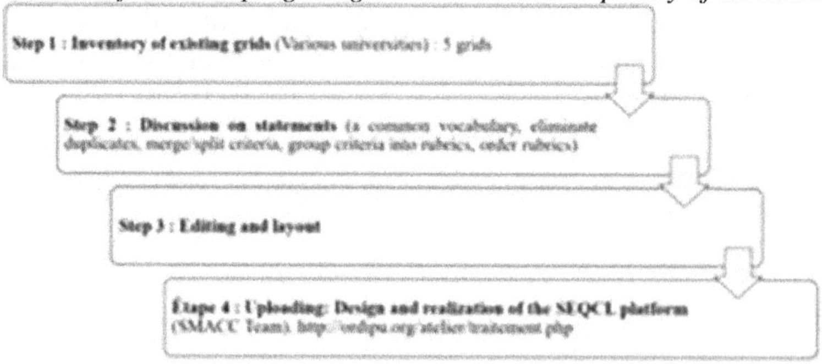

These different steps (inventory of the existing grids, grouping criteria in rubrics, editing and uploading) allow us to develop our platform SEQCL as it appears in Figure 2. SEQCL is a fully online application with no software to set up and easily usable. It will allow the teacher to evaluate the quality of an online course.

Figure 2. Screenshot of SEQCL platform

The SEQCL platform groups the quality items to design an online course, in five rubrics including 30 items (Figure 3):

- Subject and context of training,
- Educational aspects,
- Tutoring (trainer tools),
- Methods of evaluation,
- Technical aspects.

Figure 3. Screenshot of the five rubrics of SEQCL

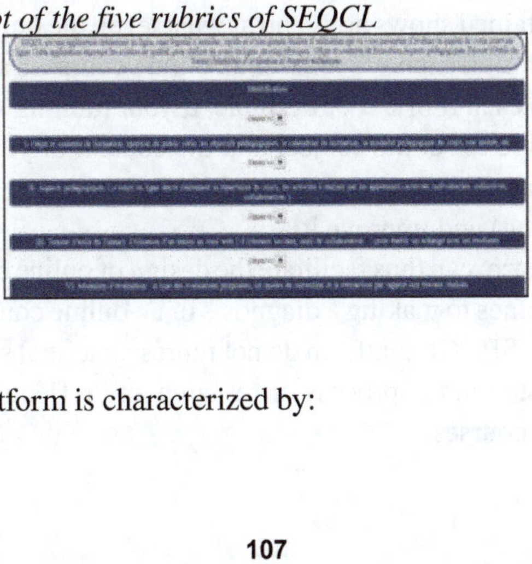

The SEQCL platform is characterized by:

- Continuity: it is a process of improving the quality of online courses.
- Flexible: it is a summarizing tool regrouping the quality items for the design of the online courses in five rubrics.
- Self-evaluative: the examination is carried out by the designer-editor of the online course or the quality insurer as part of an eLearning system.

The SEQCL platform is not:

- An evaluation of teachers but evaluate rather the quality of the course.
- A pass / fail test, but rather a diagnostic tool.
- About creating a "perfect" online course but creating an effective course whose quality is above average 80%.

To facilitate the calculation of the overall score for these items (see Figure 4), we have converted the rubrics into a dynamic platform. This open-access platform is intended for teachers who design online courses, as well as for e-learning quality assurance specialists wishing to easily verify whether an online course meets both pedagogical and technical quality criteria.

Figure 4. Screenshot of the results display of the SEQCL platform

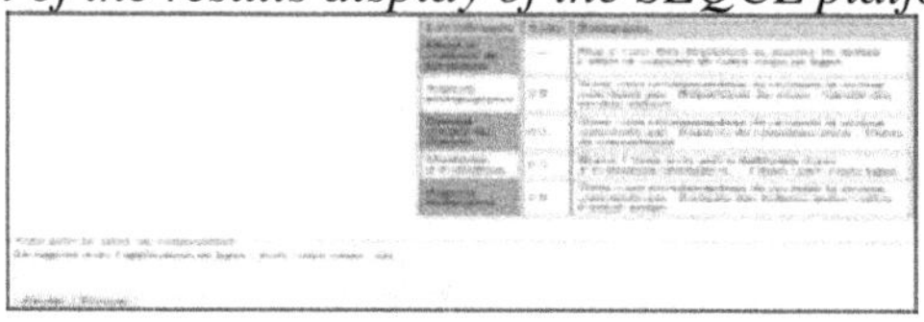

The SEQCL platform therefore leads the teacher to think about his answers, and to explain his methods relating to the educational aspects and the technical aspects of an online course. An overall score is calculated then from the participants' answers. The overall score obtained shows how, and at what level, the online course responds to quality criteria. This efficiency score for an online course is also balanced by the number of criteria in each rubric. For example, if your training score is greater than 50%, you are now able to set the subject and the context of your online course. If your score is less than 50%, it is recommended that you check the rubric (subject and context of training) and improve it.

The SEQCL platform can thus facilitate the design of online courses for teachers, especially when it comes to making a diagnosis of an online course available. However, the items on the SEQCL platform do not represent a final step. They are rather corresponding to a step in the process of research and reflection on effectiveness conditions of online courses.

2.2 Statistical Survey

Out of the 22 newly recruited teachers, only 15 responded to the items on the SE-QCL platform. As illustrated in Table 2:

Table 2. Online course titles and targeted population

Online course titles	Targeted population
Vector spaces	1st year, S2
Scientific research	Master, S1
Retro-alveolar structure	1st year
Crown casting	2nd year
Project management	Engineering curriculum
Social responsibility	Engineering curriculum
Business Management	Engineering curriculum
Management	Engineering curriculum
Communication	Engineering curriculum
Mechanics of solids	Engineering curriculum
Laws in Electro kinetics	1st year, S1
Typology of texts	1st year, S1
Liquid meningitis	1st year
Health literature search	1st year

2.3 Sample and Population

It should be noted that the population profiles of new teachers are quite varied: some of them are just beginning their university career while others have several years of experience in other institutions. The results of the survey distributed at the end of the workshop showed that teachers were generally very satisfied with the quality system. However, as specified by some authors (Bailly et al., 2015): "These questionnaires measure immediate satisfaction, but beyond that there are other more serious questions that need to be seized: What are the effects of these trainings on the actual practice of teachers?"

We have read the comments in the SEQCL platform in order to identify what is shared between the new teachers. This allowed us to notice that teachers found some difficulties in answering questions in the platform since certain criteria mentioned in the platform were not seen at the workshop of monitoring the newly recruited teachers. Of course, we present here the analysis of the main criteria concerning both the educational and the technical aspects:

- The Course must contain digital resources (videos, document ...),
- The teacher must consider spaces for communication (forum, social networks, etc.) to discuss the course content,
- The teacher must specify how students' work should be corrected.

3 FINDINGS

The statistical results show that 80% of newly recruited teachers who have filled in the online grid enrich the course content online through digital resources that exist on the internet. So, according the figure above (see Figure 5), it can be noted that the new teachers have designed online courses using resources taken mostly from YouTube. We can conclude that there is a need to schedule workshops to design pedagogical videos for teachers.

Figure 5. Online course contains digital resources.

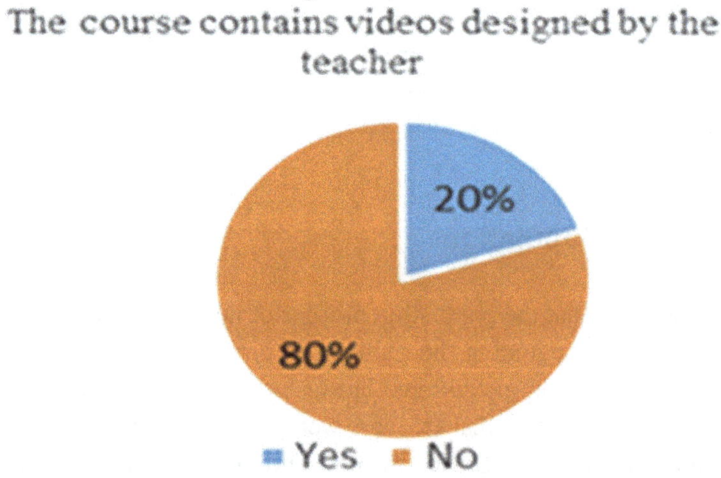

The results in the figure below (see Figure 6) show that 86.7% of teachers who answered the survey considered communication spaces (forum, social networks, etc.) to discuss the content of the course and promote social learning. However, 73% did not define a method of correcting students' work, so after reading the comments left in the SEQCL platform, it can be said that teachers have warned that the online courses will be finalised only at the beginning of the current school season and will be enriched by other digital resources. One of the teachers said, "My online course is not yet complete, there is not enough evaluation activity, it will be at the beginning of next school year. But I would say that I am satisfied of this first experience that I intend to enrich and improve further, using other tools of the Moodle platform ".

Figure 6. Communication channels and grading of student work

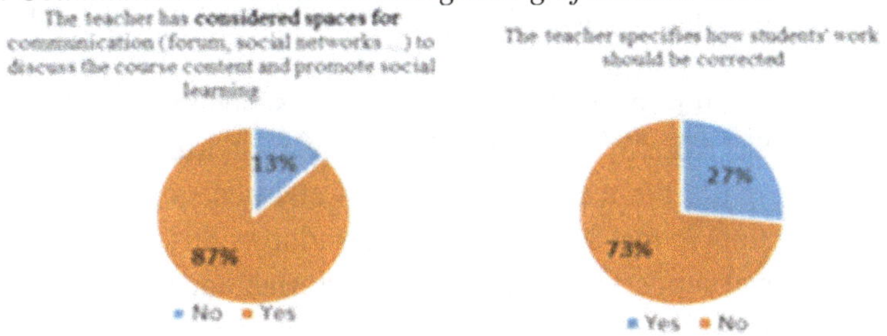

4 RESEARCH OUTCOMES

The analysis of the comments provided by teachers who used the SEQCL platform revealed both its pedagogical value and a number of technical and instructional limitations. While the platform effectively raised awareness among instructors about essential quality criteria for online course design, it also presented several constraints that merit consideration for future improvements.

From an educational standpoint, many teachers appreciated the platform's structured guidance on key aspects such as content organization, instructional strategies, and digital resource integration. However, some participants noted that the platform mainly supports a "top-down" approach to online course design, emphasizing the provision of learning aids (e.g., texts, videos, downloadable resources) while giving less attention to learner-centered and participatory methods, such as collaborative learning and peer assessment. As one teacher highlighted, "In the online course that I have designed, students have to propose a new activity to address a specific learning difficulty. This activity is then presented to peers, who work in pairs to provide constructive feedback." Such practices exemplify the type of pedagogical interactivity that is underrepresented in the current version of SEQCL.

On the technical side, two main limitations emerged. First, the platform does not generate visual reports or data visualizations, which could be helpful for both teachers and administrators in interpreting evaluation results. A simple graphical representation of scores across the different rubrics could significantly enhance usability and facilitate informed decision-making. Second, the response format is limited to binary options ("yes" or "no"), which restricts the ability to provide nuanced feedback. Several teachers expressed the need for an additional response option such as "not applicable," particularly in cases where specific criteria (e.g., video content in FLV format) were irrelevant to the course under evaluation. As one participant stated, "The course I wanted to evaluate contained no videos, so the questions about video formats were not applicable. I would have liked to respond accordingly."

These findings prompt critical reflection on the scope and depth of the data collected through SEQCL. While the platform provides a useful initial diagnosis, it does not fully capture the technical challenges and pedagogical variations encountered by educators in real-world contexts. To address this, we recommend the addition of a technical support section that offers tutorials and practical solutions to common issues faced by teachers when creating online courses. Such support would help bridge the gap between digital expectations and actual teacher competencies, allowing users to progressively integrate ICTE tools at their own pace.

In light of these observations, it is evident that the SEQCL platform has strong foundational value, but it should be further enriched with additional criteria and customizable elements that reflect the diverse realities of online teaching. Enhancements in both technical functionality and pedagogical flexibility would significantly increase its relevance and effectiveness for higher education institutions.

5 FUTURE SCOPE

Although the initial usage of the SEQCL platform remains limited only 15 out of 22 newly recruited teachers completing the evaluation grid, this early adoption nevertheless demonstrates a positive level of engagement and willingness to contribute to the platform's development. Feedback collected from participants indicates a shared interest in refining the platform by incorporating additional quality criteria that better reflect the evolving needs of online teaching environments.

Moving forward, the continuity of this research project will focus on the progressive enrichment of the SEQCL platform, particularly through the integration of new items related to both pedagogical and technical dimensions of online course design.

On the educational side, future developments will aim to address current limitations by introducing evaluation criteria that promote:

- Collaborative learning activities between students,
- Peer-to-peer assessment mechanisms to encourage student autonomy and reflection,

- And the construction of meaning through project-based learning.

These additions will support a more learner-centered pedagogy, shifting away from solely top-down instructional approaches and enabling more interactive, participative online learning experiences. The ultimate goal is to foster a model that values student agency and active engagement in their own learning process.

Regarding the technical enhancements, one of the key priorities will be the inclusion of a third response option, "Not applicable" for each evaluation item. This will address an important limitation identified during the initial testing phase, where several users indicated that binary responses ("Yes"/"No") were insufficient for representing their specific teaching contexts. For example, certain items related to video formats or specific tools may not be relevant to all types of courses, and instructors should be allowed to reflect that accurately in their evaluations.

Moreover, we envision future updates to include:

- Automated visual reporting tools, offering graphical summaries of course quality scores,
- An integrated tutorial or help section, guiding users on how to interpret results and implement improvements based on feedback,
- And potentially a customizable rubric builder, enabling institutions or instructors to adapt the platform's criteria to their own educational policies or pedagogical models.

These planned improvements aim to transform SEQCL into a more flexible, scalable, and teacher-friendly platform, capable of supporting the diverse realities of online education across disciplines and institutions. As such, the platform holds strong potential for future large-scale deployment and deeper integration within institutional quality assurance processes.

6 CONCLUSION

In the first part of this study, we presented the rationale and methodology behind the development of the SEQCL platform, an innovative tool designed to evaluate the quality of online courses based on pedagogical and technical criteria. This initial phase focused on identifying and integrating key quality standards into a user-friendly digital interface aimed at supporting educators in self-assessing their course design practices.

In the second, more practical part of the study, we analyzed feedback and data collected through actual usage of the SEQCL platform by newly recruited university teachers. This analysis enabled us to highlight both the strengths and limitations of the platform. On one hand, SEQCL succeeded in guiding educators to reflect critically on aspects such as scenario planning, digital resource integration, evaluation methods, and learner interaction. On the other hand, it also revealed areas for improvement particularly in fostering collaborative pedagogies and offering more nuanced technical flexibility.

Overall, this article has contributed to clarifying the essential standards that should guide the design, evaluation, and continuous improvement of online courses in higher education. The SEQCL platform has demonstrated its value as a diagnostic and reflective tool that does not assess individual teachers or institutions per se but rather focuses on the structure and quality of the learning content and environment.

Moreover, as a standardized yet adaptable evaluation grid, SEQCL has the potential to simplify and support the work of eLearning professionals, educational designers, and quality assurance managers. It provides a foundation for making evidence-based diagnoses and forecasts regarding the readiness, strengths, and improvement areas of online courses.

Given its modular structure and openness to enrichment, SEQCL lays the groundwork for further academic inquiry. Future research and testing on a broader scale may provide more detailed insights into the relative weight and relevance of each rubric within the platform, allowing for a more robust and context-sensitive evaluation model. The continued use and iterative enhancement of the SEQCL platform may thus contribute meaningfully to advancing educational innovation and quality assurance in digital learning ecosystems.

REFERENCES

Bailly, B., Demougeot-Lebel, J., & Lison, C. (2015). La formation d'enseignants universitaires nouvellement recrutés: Quelles retombées? *Revue Internationale de Pédagogie de l'Enseignement Supérieur*, *31*(3). Advance online publication. https://journals.openedition.org/ripes/990. DOI: 10.4000/ripes.990

Barrett, K. R., Bower, B. L., & Donovan, N. J. (2007). Teaching styles of community college instructors. *American Journal of Distance Education*, *21*(1), 37–49. DOI: 10.1080/08923640701298738

California State University. (2017). *CSU Quality Online Learning & Teaching (QOLT)*. https://dspace.calstate.edu/handle/10211.3/161320

Coristine, S., Russo, S., Fitzmorris, R., Beninato, P., & Rivolta, G. (2022, April). *The importance of student-teacher relationships*. In *Classroom Practice in 2022*. https://ecampusontario.pressbooks.pub/educ5202/chapter/the-importance-of-student-teacher-relationships/

Dong, Y. (2023). Teaching quality monitoring and evaluation in higher education through a big data analysis. *International Journal of Emerging Technologies in Learning*, *18*(08), 61–78. DOI: 10.3991/ijet.v18i08.39247

E-xcellence. (2012). *Quality assessment for e-learning: A benchmarking approach*. https://e-xcellencelabel.eadtu.eu/images/documents/Excellence_manual_full.pdf

Hua, X. (2021). Construction quality evaluation of teacher development schools based on improved artificial neural network. *International Journal of Emerging Technologies in Learning*, *16*(08), 205–220. DOI: 10.3991/ijet.v16i08.22129

Mamlouk, Z., & Dhaouadi, N. (2007). E-formation et développement des compétences. *La Revue des Sciences de Gestion*, *226–227*(4), 173–182. DOI: 10.1051/larsg:2007046

Matters, Q. (2017). *Specific review standards from the QM Higher Education Rubric* (7th ed.)., https://www.qualitymatters.org/sites/default/files/PDFs/StandardsfromtheQMHigherEducationRubric.pdf

Moore, J. C., & Shelton, K. (2013). Social and student engagement and support: The Sloan-C quality scorecard for the administration of online programs. *Online Learning : the Official Journal of the Online Learning Consortium*, *17*(1). Advance online publication. DOI: 10.24059/olj.v17i1.342

Ning, G. (2021). Evaluation criteria for quality education of physical education lessons based on logical analysis. *International Journal of Emerging Technologies in Learning*, *16*(21), 87–99. DOI: 10.3991/ijet.v16i21.26867

Niu, M. (2023). Quality evaluation of innovation and entrepreneurship education based on modelling of students' behavior sequences. *International Journal of Emerging Technologies in Learning*, *18*(09), 17–31. DOI: 10.3991/ijet.v18i09.40227

Université Laval. (2016). *Guide des bonnes pratiques de l'enseignement en ligne: Partage d'idées, conseils et suggestions*. https://www.enseigner.ulaval.ca/guide-web/guide-des-bonnes-pratiques-de-l-enseignement-en-ligne

Zweig, J. S., & Stafford, E. T. (2012). Training for online teachers to support student success: Themes from a survey administered to teachers in four online learning programs. *Journal of Online Learning Research*, *2*(4), 399–418.

Chapter 5
Tokenizing Educational Research:
Toward Incentive-Based and Decentralized Project Governance

Manal El Omari
https://orcid.org/0009-0005-7428-4813
Université Sidi Mohamed Ben Abdellah, Fès, Morocco

Mohamed El-Himer
https://orcid.org/0009-0001-1665-3650
Université Sidi Mohamed Ben Abdellah, Fès, Morocco

ABSTRACT

This chapter investigates tokenization as an innovative mechanism for managing educational research projects. Rooted in blockchain technology, it provides for transparent tracking, recognition, and motiving of individual contributions via fungible and non-fungible tokens. The proposed model integrates smart contracts (to automate much of the work), collaborative governance, and real-time monitoring. Case studies like BitDegree and ResearchHub illustrate how tokenization can enhance fairness, engagement, and accountability, and we address a number of ethical, technical, and pedagogical issues, too. These are important because they're about aligning educational research with its core values

INTRODUCTION

Contemporary educational research is playing out against a backdrop of massive shifts, where demands for performance, transparency, and social accountability ar-

DOI: 10.4018/979-8-3373-2975-8.ch005

en't just buzzwords. They're straight-up reshaping everything. The academic world has gotten sucked into this global competition game, with everyone chasing digital visibility, and that often means prioritizing stuff you can easily measure, even if it pushes the messy, collective efforts that actually make those results happen into the background.

Researchers these days are juggling a ton of conflicting expectations: churning out rock-solid knowledge, proving it matters to society and institutions, all while working in these scattered, hyper-connected teams where everyone's contributions are splintered yet totally interdependent. And in this whole setup, the gap between what's officially celebrated and what's really going on behind the scenes is getting painfully obvious (Lee et al., 2023).

The way we value academic work right now is obsessed with indexed publications, impact factors, and all those bibliometric scores. It basically ignores a huge chunk of what scientists actually do: coordinating projects, mediating conflicts, mentoring, peer-reviewing drafts, and all those unseen tasks that keep the whole knowledge-building machine running.

At the same time, universities and funding bodies are pushing harder for detailed tracking of who did what, both individually and as teams, mainly to justify the money they're pouring in and to amp up scientific accountability. But the tools they're using, like shared spreadsheets, meeting minutes, or those endless progress reports, just aren't cutting it. They're all about the final deliverables and completely miss the informal chats, the tiny cognitive leaps, the organizational tweaks, and the relational glue that holds collaborative research together. By fixating on the shiny end products, we end up hiding a massive part of the intangible work that fuels daily research life (Lamberty et al., 2020).

This blind spot is what the literature calls "invisible contributions." That's all those intellectual, organizational, and relational efforts that evaluation systems barely register, yet they're absolutely crucial for projects to succeed (Lee et al., 2023). Think mentoring students, coordinating tasks, proofreading papers, or fostering learning communities. These things often fall to junior roles or unofficial positions, and structurally, they get undervalued even though they're the backbone of quality, continuity, and team cohesion in scientific work.

So the big question becomes: how do we actually make these contributions visible, acknowledge them properly, and organize them in educational research, beyond just chasing performance metrics? This isn't only about ethics and giving credit where it's due. It's tied to the core organizational and motivational challenges in science. Without fair ways to value this stuff, collaboration fizzles, people burn out, and long-term projects fall apart.

That's where tokenization comes in, borrowed from digital ecosystems. It's basically creating and assigning digital tokens to document, recognize, and track contributions. These aren't always about money. They can just be proofs of participation, badges of appreciation, or even voting rights in collective governance. By letting us pinpoint and certify every input transparently, tokenization shines a light on those hidden efforts and builds real trust in scientific teams (Bakos & Halaburda, 2022).

There's already solid empirical work in education showing how token-based recognition boosts motivation, like in token economies that ramp up engagement, responsibility, and drive among educators (Mazzardo, 2020; Suan, 2024). Applied to scientific research, this could finally give credit to overlooked stuff like scientific mentoring, coordination, or expert sharing.

Theoretically, tokenization fits into a bigger shift in how we govern knowledge, leaning on transparency, traceability, and decentralized cooperation for validation and value (Lamberty et al., 2020). It could lead to a more inclusive scientific governance model, where work's worth isn't judged solely by final outputs but by the depth, variety, and quality of everyone's inputs.

Ultimately, tokenization isn't just a tech trick. It's a full paradigm shift in scientific recognition. By turning solo production into a traceable, valued collective process, it paves the way for a genuine "contribution economy" built on transparency, fairness, and shared responsibility. As Lee et al. (2023) point out, this could democratize academic credit by uplifting those vital but hidden efforts.

That's the ambition driving this chapter: unpacking the flaws in current recognition practices, fleshing out the idea of invisible contributions, exploring tokenization as a tool for motivation and organization, and laying groundwork for a traceability and valuation model in educational contributions. In the end, tokenization feels like a genuinely innovative lever to align ethical recognition, scientific transparency, and collaborative drive, pointing toward a brighter future for educational research.

1. THEORETICAL AND TECHNOLOGICAL FOUNDATIONS

1.1 Tokens, Blockchain, and Smart Contracts

Blockchain is this tech built on a distributed ledger (Distributed Ledger Technology, or DLT). It pretty much guarantees decentralization, immutability, transparency, and data security through that cryptographic chaining of blocks. By ditching the need for a central trusted authority, it's slowly making its way into education to lock down digital identities, certifications, and the whole traceability of learning paths (Shmatko et al., 2021).

One of the big deals with blockchain is the digital token. That's a programmable unit on the chain that stands for some kind of value, action, or contribution. Tokenization is basically turning intangible stuff (like participation, academic output, mentoring, whatever) into trackable digital assets. This setup finally makes those tiny, often overlooked micro-contributions visible in traditional education systems (Shmatko et al., 2021).

Tokens mainly split into two camps. Fungible ones (Fungible Tokens) are all interchangeable, kinda like regular money where one unit is the same as any other. Then you've got non-fungible tokens (Non-Fungible Tokens, or NFTs), which are one-of-a-kind and tied to metadata that proves authenticity and uniqueness for things like certificates, portfolios, or publications. NFTs usually run on the ERC-721 standard (Ethereum Request for Comments 721), and they really beef up document integrity while cutting down on academic fraud (Lim & Kamarudin, 2023). People are also using them for issuing those verifiable public micro-credentials.

The automation side comes from smart contracts. These are self-executing programs stored on the blockchain that kick in automatically when certain conditions are met. In education, a smart contract could just validate someone's completion of a course and spit out a diploma, no human needed, no mistakes, no middlemen (Gunawan et al., 2021). Folks usually code these in languages like Solidity, the go-to for Ethereum.

A newer twist is soulbound tokens (SoulBound Tokens, SBTs). Unlike regular NFTs, you can't trade or sell SBTs. They're designed to lock inalienably certify identity, skills, or academic rep in a way that's permanent and non-speculative, making them perfect for educational and professional credentials (Pericàs-Gornals et al., 2024).

In a nutshell, blockchain provides the secure, decentralized backbone. Tokens turn contributions and rights into tangible things. Smart contracts handle the automatic execution. When they all come together, you get this emerging model for education that's decentralized, transparent, and fair, all based on tracking actions and genuinely recognizing academic inputs.

1.2 Motivation: Self-Determination Theory (SDT)

The Self-Determination Theory (SDT), cooked up by Deci and Ryan back in 1985, is hands down one of the heaviest hitters when it comes to figuring out motivation in educational settings. It all boils down to this: people get truly engaged when three core psychological needs are met. Autonomy (feeling like you're calling the shots), competence (knowing you're actually good at what you're doing), and relatedness (feeling tight with others). SDT also splits motivation into two big camps. Intrinsic stuff, where you do it because it's fun or interesting on its own, and extrinsic, driven by outside rewards or social props. These aren't enemies though. They can totally mix and even boost each other if balanced right (Deci et al., 1991; Vitai & Benke, 2021).

One killer concept here is the overjustification effect. Basically, if you slap too many rewards on something someone already loves doing for its own sake, that intrinsic spark can fizzle out because now it feels like they're just chasing the prize. When a task starts feeling controlled or forced from the outside, people's sense of self-determination takes a hit (Deci et al., 1991).

This gets super relevant when we're talking tokenization in education setups. You know, handing out digital tokens to acknowledge contributions or achievements. The trick is designing those tokens not as straight-up transactional bribes, but as symbolic nods that celebrate the effort, the input, the commitment. If they back up autonomy, competence, and those social connections, they amp up intrinsic motivation instead of steamrolling it (Nguyen et al., 2023).

This whole angle lines up perfectly with meaningful gamification. That's the kind that focuses on real purpose, personal growth, and teamwork, not just racking up points or shallow rivalries. Gamification that goes overboard with leaderboards or fat rewards can lock people into controlled motivation and tank long-term buy-in. But if you build it around the deeper social and symbolic value of what people contribute, you get way more genuine involvement (Tan, 2018).

Digital learning spaces back this up big time. Social recognition and community vibes, especially when powered by traceable tech like blockchain for validation, crank up that sense of belonging, collective drive, and personal motivation. Recent studies straight-up confirm that solid peer interactions and shout-outs from the group juice cognitive engagement and actual learning outcomes (Deci et al., 1991; Nguyen et al., 2023).

Ultimately, SDT gives us a rock-solid lens for building human-centered tokenization in education. The goal isn't to let tech hijack motivation. It's to use tokens as genuine recognition tools that spotlight contributions without turning everything into some consumerist or cutthroat game. When tokenization feeds autonomy, competence, and social bonds, it turns into a real engine for lasting engagement, totally in sync with how human motivation actually works deep down.

1.3 Project Management and Traceability

Project management in educational research is shifting more and more toward tracking those micro-deliverables. You know, all the fragmented contributions that fly under the radar but are absolutely vital for collaborative projects to actually work. Stuff like proofreading drafts, coordinating schedules, running meetings, methodological guidance, or straight-up mentoring. These are crucial, yet classic institutional setups barely track or value them at all. They obsess over the final output and totally ignore the intermediate steps. This lack of documentation hits hard on team motivation, coordination, and accountability. That's where blockchain comes in, built on a decentralized distributed ledger (Distributed Ledger Technology, DLT).

It provides a rock-solid setup to record, timestamp, and certify every single action in a secure, unchangeable way (Heizelman & Bittner, 2024).

The digital token, this unit stamped on the blockchain, acts as auditable proof of action. It doesn't just confirm a task got done. It spells out what it was and who did it. Unlike those old-school manual methods, think reports, spreadsheets, or meeting minutes, tokens deliver automated, objective, tamper-proof verification of individual and group inputs. Heizelman & Bittner (2024) show how tossing learning tokens into pedagogical projects levels up tracking quality, ramps up responsibility, and structures collaboration way better. That opens the door to adapting it for research settings.

Tokens pull double duty too. They generate real-time activity metrics and handle automated reporting through self-executing smart contracts. These are just programs baked into the blockchain that trigger when preset conditions hit. As soon as a task gets validated, boom, the system logs the contribution and mints a token. That's unbreakable digital proof. This setup slashes the admin burden of admin tracking while making the data way more reliable (Silva et al., 2022).

Beyond just monitoring, blockchain brings built-in auditability that fits perfectly with the transparency and compliance demands in education and science worlds. Contributions aren't merely logged anymore. They become verifiable, time-stamped in order, and certifiable. This flips the script from declarative audits to real-time auditing, cutting down on errors, oversights, or any funny business with documents (Silva et al., 2022). It's a big deal for funded projects, where regularly justifying contributions and resources is key to keeping institutional trust.

Recent stuff in digital learning backs this up big time. Mixing in non-fungible tokens (Non-Fungible Tokens, NFTs) with smart contracts in immersive spots like educational metaverses lets you certify learning paths for real, auto-validate skills, and make everything interoperable across institutions (Golam et al., 2024). Every pedagogical interaction turns traceable, verifiable, and locked into a shared secure ledger. That beefs up the reliability of recognition and certification systems.

On a bigger systemic level, blockchain is reshaping how we govern educational projects. It pushes a model based on digital proof, collaborative transparency, and collective accountability. Every player, teachers, students, researchers, or partner institutions, feeds into a shared action history. That ditches centralized, declarative reporting for decentralized, verifiable setups. The structural shift builds trust between everyone, smooths out coordination, and makes recognition of contributions fairer (Domenicale et al., 2024).

So tokenization isn't just some tracking tool. It turns micro-deliverables into traceable, certifiable, auditable units of value. Tokens become go-to instruments for coordination, validation, and accountability. They make possible a fresh take on project management, rooted in transparency, verifiability, and distributed trust. This framework finally lets even the invisible or in-between contributions count fully toward measuring performance, collaboration, and the scientific value we create.

2. CASE STUDIES

2.1 BitDegree (Lithuania) – Learn-to-Earn

BitDegree, a Lithuanian platform launched back in 2017, pretty much nails what tokenization can do in education. They mash up blockchain tech with online learning setups to get people actually finishing courses by dangling real rewards in front of them, digital tokens you can swap for stuff like discounts, scholarships, or extra educational perks. The idea is dead simple: the harder you grind through the material, the more tokens land in your wallet. It's all about making education feel less like a chore and more like something worth sticking with, especially for folks who might not otherwise have the cash or support to keep going (Sparer, 2024).

The whole Learn-to-Earn thing runs on incentives. Every little step counts, whether you're watching a lecture, nailing a quiz, or jumping into a group project. Each action ties to a smart contract sitting on the blockchain. Finish a module? Boom, the contract fires off and drops a token straight into your digital wallet. No middleman, no funny business, just pure transparency and fairness baked in. Those tokens aren't only about cash either. They double as proof of how far you've come and boost your rep inside the community (Khan et al., 2022).

Gamification is the secret sauce here. Users get leaderboards, progress bars, the works. You can see how you stack up against others and trade tokens in this open educational marketplace. Recent studies back this up: when people feel like they're leveling up and actually getting good at something, their drive skyrockets. That lines up with self-determination theory, competence and accomplishment being two huge psychological needs (Fernandes & Werner, 2023). Caveat, though: if the rewards feel too much like bribes, you risk killing the inner spark that made someone want to learn in the first place. That's the overjustification effect in action.

The upsides are hard to ignore. BitDegree cranks up course completion rates, which usually suck in regular online classes. It ties effort directly to recognition, and that hits different for students from rough backgrounds or without much institutional backup. The decentralized setup kills off favoritism and builds trust between learners and platforms. Plus, it flips the script from just consuming content to actually contributing and getting paid for it (Karat & Kannimoola, 2023).

That said, it's not perfect. Leaning too hard on external rewards can backfire long-term. People might start chasing tokens instead of knowledge. And let's be real: crypto prices bounce around like crazy, so the value of those tokens isn't exactly stable. Sparer's 2024 study points out that while the initial buzz from token drops pumps engagement, you've got to pair it with solid pedagogy that actually makes skills feel meaningful. Otherwise, it turns into a vending machine instead of a classroom.

Bottom line, BitDegree shows how tokenization can shake up motivation in education. It's a legit proof-of-concept for blockchain-powered learning platforms. But it also reminds us that tech alone won't cut it. You still need autonomy, competence, and a sense of belonging, the psychological bedrock of real learning.

2.2 ResearchHub & ResearchCoin

Brian Armstrong, the guy running Coinbase, kicked off ResearchHub as this fresh take on "Research-to-Earn." The core idea is simple: every bit of scientific brainpower gets rewarded with digital tokens called ResearchCoin (RSC). We're not just talking about dropping a paper and calling it a day. You earn for peer reviews, thoughtful comments, lit reviews, data dumps, the whole nine yards. The platform wants to fix the broken incentives in academia, where it's all about pumping out papers or watching your career tank ("publish or perish"). Instead, it builds a space where every contribution gets noticed, tracked, and paid off (Lee et al., 2023).

The Research-to-Earn setup runs on decentralized blockchain logic. Your intellectual moves get logged on a public ledger and rubber-stamped by the community. Write a killer summary? Answer a tricky question? Jump into a debate? A smart contract fires up and spits out ResearchCoin automatically. These tokens double as proof you showed up and as the platform's internal cash. Use them for premium perks or to back someone else's work. Value here isn't handed down by some editor or review panel. The crowd decides (Bless et al., 2024).

One huge win with ResearchHub is how precisely it credits intellectual effort. Old-school publishing slaps names on a paper and calls it even, hiding who did what. Tokenization breaks that down. Every task, whether it's deep thinking or logistical grunt work, gets its own slice of credit. Invisible stuff like editing drafts, bridging disputes, or explaining complex ideas to outsiders finally shows up on the radar. Plus, ResearchCoin lives on a public blockchain. Every token ties back to a specific action, so the trail is crystal clear and tough to fake (Ducrée et al., 2021).

That said, the model isn't perfect. Community validation leans on votes and likes from other researchers, and that can slide into social-media popularity contests. Flashy or crowd-pleasing contributions rake in the rewards, while dense, groundbreaking, or cross-field work gets ignored. It's basically a "scientific like" button, amping up reputation bias and pushing people to chase clout instead of real insight. This echoes the Matthew effect, where the rich get richer in academic rewards (Kosmarski & Gordiychuk, 2020).

Then there's the money angle. ResearchCoin can trade on secondary markets, which opens the door to speculation. Researchers might start gaming the system for profit, ditching rigor for quick gains. That clash between cash value and knowledge value is the big hurdle for Research-to-Earn platforms. How do you spark participation without turning science into a money grab?

Key takeaway, ResearchHub and ResearchCoin are gunning to overhaul how we govern scientific work, betting on transparency, broad participation, and spread-out recognition. Nail the balance between rewards and ethics, and it could spark a fairer way to value knowledge, spotlighting the full range of brainwork. But success hinges on dodging cutthroat vibes and keeping researchers hooked on the joy of discovery, the real engine of creativity and breakthroughs.

2.3 Gitcoin: Collaborative Governance in Action

Gitcoin stands out as this killer real-world experiment in decentralized, community-driven governance for open source stuff. Kicked off back in 2017, it runs on crowdfunding fueled by a Decentralized Autonomous Organization (DAO). Basically, decisions get made collectively by token holders, no central boss calling the shots. The DAO (Decentralized Autonomous Organization) operates through smart contracts locked on the Ethereum blockchain, which locks in transparency, traceability, and automatic rule enforcement. The whole point is to fund open source projects while spreading resources fairly based on participation and peer recognition (Feichtinger et al., 2023).

At its core, Gitcoin revolves around quadratic funding, this clever math trick cooked up by Vitalik Buterin, Glen Weyl, and Zoe Hitzig to crank up the collective punch of donations. A project's funding doesn't just hinge on the total cash pile. It also factors in how many unique contributors chip in, putting a premium on broad community backing. In practice, collected funds get topped up with a matching grant that scales with contribution spread, making everything more democratic and representative. This setup pushes for fairer resource flow in the open source world, dialing back the sway of big-money players (Pasquini, 2020).

Gitcoin's community governance thrives on active member involvement in decisions, proposal reviews, and fund handling. Holders of GTC tokens (the platform's native currency) can pitch ideas and vote on project strategy, distribution rules, or which programs get the green light. This approach aims to boost transparency and shared accountability while keeping open source projects sustainable. That said, some empirical studies flag real limitations. Even with all the decentralization hype, voting power still clusters with a handful of heavy hitters, raising doubts about true flat governance (Feichtinger et al., 2023).

Digging into how Gitcoin actually works reveals what makes projects click: clear descriptions, project length, reward size, and the problem's nature. One big empirical dive into over 4,000 Gitcoin projects shows that solid write-ups and task coherence massively boost success odds, highlighting structured collaboration's starring role in open source ecosystems (Choetkiertikul et al., 2023). Bottom line, decentralized governance doesn't ditch human coordination. It just provides a digital backbone, still relying heavily on quality dialogue and community habits.

Gitcoin's appeal now stretches way beyond software engineering. Its core ideas (transparency, participation, and decentralized collective funding) could totally reshape education, especially research and training. Picture an educational DAO where teachers, researchers, and students jointly decide resource allocation, back innovative teaching projects, or reward hidden gems like science mediation or tutoring. Gitcoin serves up real inspiration for collaborative education governance, swapping institutional hierarchies for distributed engagement and group incentives (Küng et al., 2025).

Wrapping up, Gitcoin proves blockchain can power fresh cooperation models built on distributed trust, open governance, and community buy-in. Sure, it's got room for improvement, especially around decision power concentration, but it sketches a game-changing future for how scientific and educational communities handle resources and credit contributions down the line.

2.4 Cross-Model Breakdown

When you stack up BitDegree, ResearchHub, and Gitcoin side by side, a few big ideas keep popping up across the board: traceability, incentives, and recognition that's spread out instead of top-down. In all three setups, the blockchain works like this giant, open ledger that nobody controls alone. It makes sure every contribution (whether it's teaching stuff, pushing science forward, or just helping the community) stays transparent and easy to check. The whole token thing brings in rewards you can actually measure, which gets people more involved and gives proper credit to what each person brings to the table. Basically, these platforms flip the script on what "counts." Value doesn't come from your title or a fancy journal anymore; it's about what you actually do for the group (Lee et al., 2023; Khan et al., 2022; Feichtinger et al., 2023).

That said, they've all got the same weak spots baked in. Everything runs on token economics, so there's a real chance the feel-good recognition gets overshadowed by cold hard cash value. That financial angle can kill off the inner drive people had in the first place and spark all kinds of speculative games. Plus, even though decentralized governance sounds fair and open, power still clumps up in weird ways, and popularity contests sneak in through the back door. Put it all together, and you see why we need solid ethical rules and clear regulations to keep incentives, fairness, and straight-up integrity in check whenever tokenization hits education or research (Ducrée et al., 2021).

3. PROPOSED TOKENIZED MANAGEMENT MODEL

3.1 Token Architecture and Typology

A permissioned blockchain fits academic ecosystems like a glove when you need collective governance, ironclad security, and bulletproof traceability. Unlike wide-open public chains such as Bitcoin or Ethereum, this setup limits access to a vetted consortium of players: universities, research institutes, labs, funding agencies. Each runs its own node, basically a server that helps validate and log every transaction. The result is distributed validation without leaking sensitive data, fully in line with the EU's General Data Protection Regulation (GDPR), which demands top-tier privacy safeguards. Institutions jointly guard the integrity of the ledger, ditch any single point of control, and still deliver institutional transparency that plays nice with legal frameworks (Ra et al., 2020).

On top of this backbone sits a typology of digital tokens crafted to mirror the full spectrum of scientific contributions.

First up are recognition tokens, modeled after non-transferable SoulBound Tokens (SBT). You can't sell, trade, or gift them. They lock in lasting academic facts: a published paper, a certification, an institutional duty, or a personal milestone. Their permanence delivers symbolic credit that sticks around forever and sidesteps any whiff of financial speculation. Recent studies show these tokens enable rock-solid, tamper-proof cross-institution authentication while respecting GDPR privacy rules (Boi et al., 2024).

Next come contribution tokens, fungible by design, meaning they're interchangeable and stackable. They quantify measurable chunks of academic labor: deliverables, teaching modules, technical tasks, structured peer reviews, or scientific writing. Weighting happens through criteria co-set by academic governance: complexity, scientific reach, cognitive load, collective impact, or interdisciplinarity. These tokens feed straight into automated smart contracts that calculate, verify, and archive contributions under crystal-clear rules. The multi-layer setup, handling protocols, transactions, and wallets, ensures interoperability, scalability, and end-to-end traceability (Lesavre et al., 2021).

Third are engagement tokens, focused not on output but on sustained interaction: showing up to meetings, coordinating efforts, giving evaluative feedback, co-hosting events, or fueling practice-community conversations. Unlike contribution tokens, they reward continuity rather than results, bolstering social cohesion and spotlighting invisible work that traditional systems routinely undervalue (Goldston et al., 2023).

Tech-wise, the platform rides a Layer 2 infrastructure built atop a primary Layer 1 chain to slash computational costs and turbocharge processing speeds. Layer 2 solutions can cut gas fees by up to 90% while preserving cryptographic security via cross-chain validation mechanisms (Bayan & Banach, 2024). This keeps the economics sane and makes large-scale institutional rollout realistic.

Legal compliance leans on pseudonymization, swapping direct identifiers for cryptographic ones, plus zero-knowledge proofs that let you prove something without spilling the details, delivering traceability without compromising privacy (Fekete & Kiss, 2023).

Governance stays distributed: every consortium member validates transactions and steers the smart contracts, dodging centralization pitfalls. By splitting recognition, contribution, and engagement, the system captures the whole participation landscape. Marrying a permissioned blockchain, Layer 2 efficiency, and GDPR compliance yields a tough, transparent, resilient, cooperative model where academic value isn't just measured; it's certified, legitimized, and shared.

3.2 Roles, Features, and Pseudo-Code

Bringing a tokenized management system into educational research means shaking up how roles work, how decisions get made, and how credit is handed out. The whole setup borrows heavily from Decentralized Autonomous Organizations (DAOs). You know, those distributed governance systems where rules, decisions, and recognition aren't bottled up in one central office but get hashed out collectively and baked into a transparent protocol. Here, automation and traceability don't kick human responsibility to the curb. They actually beef it up in this hybrid model that mixes algorithmic rules with ethical oversight (Feichtinger et al., 2023).

The system revolves around four main player types: researchers, coordinators, trainees, and funders. Researchers aren't just cranking out final papers. They dive into all the in-between stuff like coordinating tasks, peer-reviewing drafts, mentoring, or hammering out methodologies. Every validated contribution mints a Jeton_Contribution. That's a digital unit logged on the blockchain, certifying the action and making it visible and verifiable to the entire academic consortium. This setup cuts down on subjective credit-grabbing and ramps up fairness in scientific evaluation (Lee et al., 2023).

Coordinators act as the bridge between people and tech. They check contributions for compliance before green-lighting them, ensuring scientific quality, relevance, and alignment with project goals. Think of them like curators in DAOs. They can suggest tweaks to governance rules, such as reshuffling token weights, adding new metrics, or fine-tuning validation criteria. This role blends science, organization, and ethics. It keeps the system flexible while locking in rigor and transparency for decision-making.

Trainees bring the learning and training angle. Their input goes way beyond grunt work. It fits into a valued learning path. Every skill or milestone validated can spawn a competency token, often as a Non-Fungible Token (NFT), tucked into a digital academic wallet. Non-transferable and personalized, it works like a badge proving a skill mastered or a task nailed, all in the spirit of lifelong learning and pedagogical recognition (Khan et al., 2022).

Funders stay out of day-to-day governance but serve as institutional watchers. The system gives them real-time tracking backed by cryptographic proof instead of manual declarations or polished reports. This auto-auditability, powered by blockchain, builds trust, slashes manipulation risks, and bolsters budget justifications (Silva et al., 2022).

Three core features shape the user experience: a dashboard, a traceable history, and automated reporting. The dashboard offers an up-to-date view of tokens earned, their types, and overall progress. It boosts transparency and keeps everyone engaged. The history functions as an immutable log of interactions, linking each action to an author, validator, timestamp, and unique ID. Automated reporting kills redundant paperwork and ensures deliverables match actual contributions.

Token assignment follows an algorithmic logic drawn from smart contracts. Those are self-executing programs that fire when predefined conditions hit. The basic idea boils down to this pseudo-code: If Contribution_validée = True, then Assign Jeton_Contribution (+1), else no assignment. This automation wipes out arbitrariness, standardizes rules, and strengthens perceived fairness among participants.

Participatory governance hinges on token-weighted voting, where involvement level sets decision-making clout. Proposals to tweak the protocol, add criteria, or reallocate rules get debated and then etched into the blockchain. While this amps up internal democracy, it needs safeguards to prevent power from pooling with a few big players (Feichtinger et al., 2023).

Overall, the system weaves human contributions, scientific mediation, and transactional validation into a framework where recognition isn't just claimed. It's certified, traceable, and collective. By automating merit distribution while safeguarding ethical decisions, it creates a setup that doesn't sideline humans. It puts them front and center in a more transparent, equitable, and high-performing governance for educational research.

4. BENEFITS AND DRAWBACKS

4.1 Benefits

Bringing blockchain and tokenization into educational and research settings packs a serious punch when it comes to structural and teaching perks. It's basically a game-changer for how things get organized. The big win right off the bat is transparency. Blockchain is this distributed ledger that nobody can tamper with, so every single move gets stamped with a time, checked by anyone, and seen by the whole network. That kills off the murky corners you usually find in old-school top-down systems, especially around giving credit, handing out funding, or tracking who did what. In academic projects, it means you can follow every single contribution like a breadcrumb trail, making sure everyone's brainpower and tech skills get a fair shake. Morales-Alarcón and crew (2024) point out that this setup beefs up trust and keeps things honest in scientific evaluations by pretty much shutting down any chance of fudging data, playing favorites, or straight-up plagiarism thanks to those rock-solid records.

Another huge plus is the lasting motivation you get from decentralized recognition. Unlike the usual reward systems run from the top, tokenization spreads the love fairly, spotlighting both the tiny unseen tasks and the big splashy wins. Platforms can hand out symbolic tokens as rewards, which keeps people pushing forward and feeling like they really belong. Recent studies back this up, showing that when efforts are out in the open and verifiable, folks stay happier and more driven over the long haul. It lines up perfectly with those core needs for autonomy and competence from self-determination theory (Ballandies, 2022).

Blockchain also cuts down on drama and makes sure credit lands where it should within academic circles. By putting every contribution on display, it wipes out those fuzzy arguments over who came up with what or who actually finished the job. Smart contracts bake the rules right into the code, doling out tokens automatically and without bias based on whatever criteria the group agreed on upfront. Panagiotidis (2022) talks about this algorithmic governance angle, saying it builds solid trust between institutions and keeps roles and duties crystal clear in team projects.

Then there's the automated reporting and the way data plays nice across systems. Everything about contributions lives in one central ledger, so you can spit out verifiable activity reports without breaking a sweat. That slashes paperwork while making the info way more consistent and trustworthy. Funders, project leads, and universities get real-time dashboards they can actually rely on. Samala and colleagues (2024) confirm that blockchain can shake up educational governance by streamlining certification, tracking, and transparency, plus giving everyone a standard way to swap data between schools.

All in all, mixing traceability, sustained motivation, fair recognition, and hands-off reporting turns blockchain into a powerhouse for boosting reliability, accountability, and plain old fairness in education and research worlds.

4.2 Limitations

Blockchain and tokenization bring some serious perks to educational projects, no doubt, but they also drag along a bunch of structural, psychological, and technical headaches that we really need to unpack carefully. First off, there's this thing called overjustification, something Deci and Ryan flagged way back in 1985 with their self-determination theory. The idea is pretty straightforward: pile on too many external rewards, and over time, people's inner drive takes a hit. Tokens can start feeling way too much like cold, hard incentives, pushing folks to chase the prize instead of getting lost in the sheer joy of learning or intellectual curiosity. Ballandies (2022) actually ran studies showing that crypto rewards can trigger a "crowding-out" effect, where intrinsic motivation gets shoved aside for extrinsic goals, and that tanks both the quality and the creativity of what people produce.

Another big hurdle is the digital divide, plus all the cultural pushback against jumping on the blockchain bandwagon. Tartan (2023) points out that shifting to decentralized digital systems can widen the gap between tech-savvy users and everyone else. In schools and universities, that means uneven adoption among teachers, researchers, and students, especially in places where tech resources are thin on the ground. On top of that, you've got cultural resistance rooted in distrust of decentralization or straight-up fear of losing institutional control, and that really slows down rolling these innovations out in traditional academic settings (Tartan, 2023).

Then there's the sheer technical complexity and the steep learning curve, which throw up a massive barrier to making this stuff stick long-term. Mohammad and Vargas (2022) dug into how integrating blockchain into higher education demands a solid grasp of technical protocols, cybersecurity, and digital identity management, skills that just aren't widespread in academia yet. Keeping these systems running also calls for beefy infrastructure and top-tier IT know-how, which racks up hefty implementation costs and locks you into relying on specialized vendors (Mohammad & Vargas, 2022).

Add to that the lack of clear ethical and regulatory frameworks around data ownership, privacy, and algorithmic governance, and you've got a recipe for potential misuse. Garrido (2023) warns that without rock-solid legal grounding, tokens can blur the line between symbolic recognition and actual economic value, stirring up ethical mess in how we value intellectual work (Garrido, 2023).

In summary, while blockchain holds huge promise for shaking up educational governance, rolling it out demands a cautious game plan built on proper training, ethical regulation, and balanced pedagogical design. Otherwise, the tech risks overshadowing the deeply human goals of research and learning.

5. STRATEGIC RECOMMENDATIONS

5.1 A Five-Step Roadmap

Rolling out a blockchain-powered system for tracking educational and scientific contributions calls for a slow-burn strategy. Think of it as a five-phase rollout: start with a pilot prototype, move to hands-on evaluation, lock in an ethics framework, scale up bit by bit, and cap it with shared governance. This playbook pulls straight from studies on digital-governance pilots and decentralized education consortia, which all hammer home the need for step-by-step, ethics-first progress (Anyanwu et al., 2023).

Step one is the pilot prototype. Keep it small, maybe a tight-knit university consortium, and test the core features: logging contributions, handing out tokens, spitting out automated reports. The goal here is to prove it actually works and that people don't hate it. Hallwright and Carnaby (2019) point out that blockchain pilots catch the non-tech headaches early, think pushback from staff or skimpy training, stuff that blows up later if you go big too fast (Hallwright & Carnaby, 2019).

Step two brings everyone into the loop for participatory evaluation. Researchers, coordinators, students, funders, the whole crew. You run ongoing checks on how the tools affect teamwork, motivation, and credit-sharing. Qualitative feedback becomes your tuning knob for tech tweaks and governance rules. Baharmand et al. (2021) studied blockchain in humanitarian projects and found that looping in end-users from day one keeps the solution grounded and socially palatable (Baharmand et al., 2021).

Step three is all about nailing down an ethical framework. Data privacy, algorithmic fairness, responsible governance, the works. Draw from global digital-ethics guidelines, like Shutova and Begishev (2023), who argue every bleeding-edge project needs a living code of conduct. In education, that means spelling out who owns the data, what the tokens really mean, and how to appeal if someone feels screwed. You also need an independent oversight crew to watchdog transparency, equity, and bias (Shutova & Begishev, 2023).

Step four handles gradual integration. Spread the system to new departments or partner schools while keeping everything interoperable. By now the tech is hardened (security, standardized protocols) and the paperwork is sorted (partnership deals, regulatory nods). Look at the Malaysia National Blockchain Roadmap: slow, legally backed, cross-institution scaling is what makes blockchain stick (AlAfnan & MohdZuki, 2024).

Step five flips the switch to shared governance, DAO-style. Consortium members vote on decisions through a crystal-clear mechanism. You balance tech (security, interoperability), organizational (coordination, role clarity), and legal-ethical layers. Mustafa et al. (2024) stress that governance is the make-or-break piece for decentralized systems; success hinges on blending community autonomy with institutional guardrails (Mustafa et al., 2024).

Bottom line: this roadmap is methodical, collaborative, and layered. It lets tokenization take root in education without tripping over legitimacy issues, scientific, social, or institutional.

5.2 Pilot Project Idea for an Inter-University Education Consortium

Picture this pilot kicking off inside a cross-university consortium, a bunch of schools that all vibe on the same wavelength when it comes to shaking up teaching and research. The whole point? Try out a blockchain platform that tracks and gives props to academic contributions using a system of participation tokens logged on a shared ledger.

Here's how it would roll: every time someone chips in on a research gig, co-writes a paper, mentors a newbie, or runs a conference, they earn tokens. No cash value, just bragging rights and proof of how deep they went in and how solid their work was. Those tokens could then feed into certifications, internal reviews, or even show up as performance markers in the usual institutional reports.

To keep everything above board and truly collective, the consortium would run on an educational DAO. Member schools and individual participants get to vote on stuff like token-award rules, validation criteria, and any tweaks to the setup. That spreads decision-making power around fairly and leaves a crystal-clear trail of every choice and contribution.

The group would double as an ethics lab, playing with different ways to verify stuff and govern data. Pulling ideas from frameworks laid out by Baharmand et al. (2021) and Mustafa et al. (2024), the project would wear two hats, scientific and institutional, ultimately spitting out practical guidelines on using blockchain responsibly in education.

5.3 Why We Absolutely Need an Ethics Committee and a Solid Oversight Framework

Rolling out a project like this means we have to set up an inter-institutional ethics committee right from the start. Its job would be to keep an eye on the legal, technical, and ethical sides of everything. Picture a mixed crew: researchers, lawyers, ed-tech experts, and a couple of student reps. Their main gig? Digging into what tokenization really means for academic credit and making sure it doesn't open any ethical cans of worms.

The committee would hammer out the core ethical governance rules: data confidentiality, the right to be forgotten, dodging algorithmic biases, and keeping validation criteria crystal clear. They'd also double-check that the whole operation lines up with national laws and the General Data Protection Regulation (GDPR) across Europe.

Adeusi et al. (2024) drive the point home: bake ethics and transparency into the design from day one, not as some afterthought. They push for shared accountability between people and tech systems, striking the right balance between automation and moral judgment (Adeusi et al., 2024).

Key takeaway, putting together an ethics committee and a tight oversight setup is non-negotiable. It's the only way to make sure an educational blockchain actually serves science and learning without screwing over the core values we all swear by: integrity, fairness, and collective responsibility.

CONCLUSION

Everything we've dug into in this work keeps circling back to the same core idea. Tokenization isn't just some side gimmick. It's a full-on systemic lever to shake up governance, recognition, and motivation across educational and scientific ecosystems. By blending blockchain's killer features (think traceability, transparency, immutability) with a smart incentive setup built around digital tokens, it flips the script on fairness and collective accountability. We're talking about ditching those old-school, centralized, and often murky administrative setups for something decentralized that runs on distributed trust and real community involvement. Sunyaev et al. (2021) nail it when they say tokenization rewires how we validate stuff and hand out resources, cutting out institutional middlemen in favor of clear, auditable tech protocols.

In the education space, tokenization steps up as a legit tool for empowerment and shining a spotlight on those tiny, overlooked contributions. With research and higher ed constantly under the gun to produce, justify, and measure performance, decentralized systems finally let us give props to the invisible engagement: coordinating chaos, mentoring rookies, offering cognitive backups, or just showing up in community discussions. Handing out tokens as proof of contribution amps up symbolic recognition while keeping a tight track on collective efforts. It cultivates this open, documented collab culture that lines up perfectly with open science principles.

That said, pulling this off hinges on nailing the motivation balance. Ballandies (2022) showed that leaning too hard on extrinsic incentives can trigger overjustification, where you end up killing the intrinsic drive tied to curiosity, autonomy, and meaning-making. This hits especially hard in education, where learning and research need to stay ends in themselves. The trick is designing tokenization not as some transactional reward scheme, but as symbolic acknowledgment that boosts intellectual passion instead of crowding it out.

This balance ties right into what Esmaeilzadeh and Mirzaei (2023) found. Extrinsic incentives, whether cash or badges, won't drive blockchain adoption if trust, tech literacy, and security aren't locked in. They stress that lasting motivation only kicks in when users actually get the upsides, rules, and limits of the tech. So, blockchain education becomes a must-have prerequisite for weaving it into academia without drama.

On the bigger systemic level, tokenization cracks open the door to decentralized educational governance that spreads decision-making power more evenly. Barbereau et al. (2023) draw from DeFi experiments with token-based governance, highlighting risks like power concentration and the need for stronger collective oversight. Applied to education, this drives home the point of balancing community autonomy with ethical guardrails to keep algorithmic logic from baking in existing inequalities.

All the perks we've flagged throughout (traceability, transparency, distributed recognition, automated reporting) need counterweights in ethical vigilance and solid training. Crafting an ethical and regulatory framework is non-negotiable to govern token creation, exchange, and use. Chaffer et al. (2024) argue that decentralized tech governance has to mix innovation with human oversight to dodge technocratic pitfalls and ensure shared accountability. Things like an ethics committee, data protection policies, and algorithm transparency are table stakes for blockchain to uphold education's core values: inclusion, equity, and intellectual autonomy.

Bottom line, educational tokenization isn't merely a tech upgrade. It's a cultural and institutional overhaul. It forces us to rethink how learning communities recognize, value, and self-govern their inner workings. Implemented with care, rigor, and real dialogue, it could forge a decentralized educational ecosystem grounded in transparency, co-responsibility, and distributed recognition.

In the end, tokenization feels less like a break from tradition and more like an extension of education's foundational values: cooperation, justice, and the pursuit of meaning. It opens the door to a fresh digital humanism, where tech doesn't supplant people but supercharges our ability to create together, learn as a collective, and govern fairly. The future of educational governance might just hinge on this fusion of blockchain and ethics, transparency and humanity.

REFERENCES

Adeusi, K. B., Jejeniwa, T. O., & Jejeniwa, T. O. (2024). *Advancing financial transparency and ethical governance: Innovative cost management and accountability in higher education and industry*. International Journal of Management & Entrepreneurship Research., DOI: 10.51594/ijmer.v6i5.1099

AlAfnan, M., & MohdZuki, S. F. (2024). Malaysia's National Blockchain Roadmap: A critical discourse analysis of focus, goals, and challenges. *World Journal of English Language, 14*(5), 482. Advance online publication. DOI: 10.5430/wjel.v14n5p482

Anyanwu, A., Dawodu, S. O., Omotosho, A., Akindote, O. J., & Ewuga, S. K. (2023). Review of blockchain technology in government systems: Applications and impacts in the USA. *World Journal of Advanced Research and Reviews, 20*(3), 863–875. Advance online publication. DOI: 10.30574/wjarr.2023.20.3.2553

Baharmand, H., Saeed, N., Comes, T., & Lauras, M. (2021). Developing a framework for designing humanitarian blockchain projects. *Computers in Industry, 131*, 103487. DOI: 10.1016/j.compind.2021.103487

Bakos, Y., & Halaburda, H. (2022). *Overcoming the coordination problem in new marketplaces via cryptographic tokens*. SSRN Electronic Journal. DOI: 10.2139/ssrn.4028849

Ballandies, M. (2022). To incentivize or not: Impact of blockchain-based cryptoeconomic tokens on human information sharing behavior. *IEEE Access : Practical Innovations, Open Solutions, 10*, 74111–74130. DOI: 10.1109/ACCESS.2022.3189774

Barbereau, T. J., Smethurst, R., Papageorgiou, O., Sedlmeir, J., & Fridgen, G. (2023). Decentralised finance's timocratic governance: The distribution and exercise of tokenised voting rights. *Technology in Society, 73*, 102251. Advance online publication. DOI: 10.1016/j.techsoc.2023.102251

Bayan, T., & Banach, R. (2024). *A privacy-preserving DAO model using NFT authentication for the Punishment not Reward blockchain architecture*. ArXiv. https://doi.org//arXiv.2405.13156DOI: 10.48550

Bless, C., Denzler, A., Karras, O., & Auer, S. (2024). *A reputation system for scientific contributions based on a token economy*. In *Proceedings of the 27th International Conference on Knowledge Engineering and Knowledge Management* (pp. 41–61). Springer. DOI: 10.1007/978-3-031-72437-4_3

Boi, B., Cirillo, F., De Santis, M., & Esposito, C. (2024). Soulbound tokens: Enabler for privacy-aware and decentralized authentication mechanism in medical data storage. *Blockchain in Healthcare Today*, *7*(2), 334. DOI: 10.30953/bhty.v7.334 PMID: 39649415

Chaffer, T. J., Bayo, J. G., & Gemach, O. (2024). *On the ETHOS of AI agents: An ethical technology and holistic oversight system*. ArXiv. https://doi.org//arXiv.2412 .17114DOI: 10.48550

Choetkiertikul, M., Puengmongkolchaikit, A., Chandra, P., Ragkhitwetsagul, C., Maipradit, R., Hata, H., Sunetnanta, T., & Matsumoto, K. (2023). Studying the association between Gitcoin's issues and resolving outcomes. *Journal of Systems and Software*, *206*, 111835. DOI: 10.1016/j.jss.2023.111835

Deci, E., Vallerand, R., Pelletier, L., & Ryan, R. (1991). Motivation and education: The self-determination perspective. *Educational Psychologist*, *26*(3), 325–346. DOI: 10.1207/s15326985ep2603&4_6

Deci, E. L., & Ryan, R. M. (1985). *Intrinsic motivation and self-determination in human behavior*. Springer., DOI: 10.1007/978-1-4899-2271-7

Domenicale, I., Viano, C., & Schifanella, C. (2024). Blockchain for local communities: An exploratory review of token economy aspects. *Frontiers in Blockchain*, *7*, 1426802. DOI: 10.3389/fbloc.2024.1426802

Ducrée, J., Etzrodt, M., Bartling, S., Walshe, R., Wittek, N., & Kogias, D. (2021). Unchaining collective intelligence for science, research, and technology development by blockchain-boosted community participation. *Frontiers in Blockchain*, *4*, 631648. DOI: 10.3389/fbloc.2021.631648

Esmaeilzadeh, P., & Mirzaei, T. (2023). Role of incentives in the use of blockchain-based platforms for sharing sensitive health data. *Journal of Medical Internet Research*, *25*, e41805. Advance online publication. DOI: 10.2196/41805 PMID: 37594783

Feichtinger, R., Fritsch, R., Vonlanthen, Y., & Wattenhofer, R. (2023). *The hidden shortcomings of (D)AOs: An empirical study of on-chain governance*. ArXiv. https://doi.org//arXiv.2302.12125DOI: 10.48550

Fekete, D., & Kiss, A. (2023). Toward building smart contract-based higher education systems using zero-knowledge Ethereum virtual machine. *Electronics (Basel)*, *12*(3), 664. DOI: 10.3390/electronics12030664

Fernandes, F., & Werner, C. (2023). *Towards a blockchain-based software engineering education*. ArXiv. /arXiv.2304.04549DOI: 10.5753/cblockchain.2023.230488

Garrido, J. M. (2023). *Digital tokens: A legal perspective.* IMF Working Papers. DOI: 10.5089/9798400250149.001

Golam, M., Tuli, E. A., Alief, R. N., Kim, D. S., & Lee, J. M. (2024). Meta-learning: A digital learning management framework using blockchain for metaverses. *IEEE Access : Practical Innovations, Open Solutions*, *12*, 92774–92786. DOI: 10.1109/ACCESS.2024.3408878

Goldston, J., Chaffer, T. J., Osowska, J., & von Goins, C. (2023). *Digital inheritance in Web3: A case study of Soulbound tokens and the social recovery pallet.* ArXiv. https://doi.org//arXiv.2301.11074DOI: 10.48550

Gunawan, I., Lutfiani, N., Aini, Q., Suryaman, F. M., & Sunarya, A. (2021). Smart contract innovation and blockchain-based tokenization in higher education. *Journal of Education Technology*, *5*(4), 256–264. DOI: 10.23887/jet.v5i4.40665

Hallwright, J., & Carnaby, E. (2019). Complexities of implementation: Oxfam Australia's experience in piloting blockchain. *Frontiers in Blockchain*, *2*, 10. DOI: 10.3389/fbloc.2019.00010

Heizelman, S., & Bittner, K. (2024). *Empowering engineering students: Blockchain learning tokens for assessing ABET outcomes.* ASEE., DOI: 10.18260/1-2--47245

Karat, G., & Kannimoola, J. M. (2023). *Revolutionizing education: Blockchain-based incentive systems.* ICICCS, 1370–1375. DOI: 10.1109/ICICCS56967.2023.10142632

Khan, M. F. F., Viet, P. D., & Sakamura, K. (2022). *Blockchain-based incentive management system for educational organizations.* ICECCME., DOI: 10.1109/ICECCME55909.2022.9988717

Kosmarski, A., & Gordiychuk, N. (2020). Token-curated registry in a scholarly journal. *Learned Publishing*, *33*(4), 370–379. DOI: 10.1002/leap.1302

Küng, R., & Giaglis, G. M. (2025). *Value distribution in OSS: The role of DAOs.* ICEDEG, 160–168. DOI: 10.1109/ICEDEG65568.2025.11081577

Lamberty, R., Poddey, A., Galindo, D., de Waard, D., Koelbel, T., & Kirste, D. (2020). Efficiency in digital economies: A primer on tokenomics. *Digital Economics Review*, *4*(2), 55–73.

Lee, J., Moroso, M., & Mackey, T. (2023). Unblocking recognition: A token system for academic contribution. *Frontiers in Blockchain*, *6*, 1136641. DOI: 10.3389/fbloc.2023.1136641

Lesavre, L., Varin, P., & Yaga, D. (2021). Blockchain networks: Token design and management overview. *NISTIR, 8301*. Advance online publication. DOI: 10.6028/NIST.IR.8301

Lim, J. O., & Kamarudin, D. (2023). NFTs in combatting certificate fraudulence. *IJEBT, 1*(1), 1–15. DOI: 10.59021/ijebt.v1i1.13

Mazzardo, E. (2020). *A token economy for school performance optimization*. AARN. DOI: 10.2139/ssrn.3696544

Mohammad, A., & Vargas, S. (2022). Challenges of using blockchain in education. *Applied Sciences (Basel, Switzerland), 12*(13), 6380. Advance online publication. DOI: 10.3390/app12136380

Morales-Alarcón, C. H., Bodero-Poveda, E., Villa-Yánez, H. M., & Buñay-Guisñan, P. A. (2024). *Blockchain & peer review*. Publications., DOI: 10.3390/publications12040040

Mustafa, G., Rafiq, W., Jhamat, N., Arshad, Z., & Rana, F. A. (2024). *Blockchain governance models in e-government*. IJLM., DOI: 10.1108/IJLMA-08-2023-0172

Nguyen, P.-B.-T., Degrave, P., Van Steendam, E., & Sercu, L. (2023). *Self-determination in EMI*. IJ Educational Research Open., DOI: 10.1016/j.ijedro.2023.100295

Panagiotidis, P. (2022). Blockchain in language learning. *European Journal of Education, 5*(1), 66–82. DOI: 10.26417/443gjm83

Pasquini, R. (2020). *Quadratic funding and matching funds*. Working Paper.

Pericàs-Gornals, R.. (2024). Rejectable soulbound tokens in credentials. *Annales des Télécommunications, 79*, 843–855. DOI: 10.1007/s12243-024-01032-6

Ra, G., Seo, D., Bhuiyan, M. Z. A., & Lee, I. (2020). Anonymous protocol for privacy in blockchain. *Electronics (Basel), 9*(8), 1183. DOI: 10.3390/electronics9081183

Samala, A.. (2024). Blockchain in education: Opportunities and challenges. *IJIMT, 18*(1), 20–42. DOI: 10.3991/ijim.v18i01.46307

Shmatko, O., Borova, T., Yevseiev, S., & Milov, O. (2021). Tokenization of educational assets. *ScienceRise, 1*(45), 4–10. DOI: 10.15587/2519-4984.2021.232321

Shutova, A. A., & Begishev, I. (2023). *Ethical code for medical robotics activity*. DOI: 10.21202/978-5-8399-0803-1_2023_2_15

Silva, R., Inácio, H., & Marques, R. P. (2022). Blockchain implications for auditing. *The International Journal of Digital Accounting Research, 22*, 1–25. DOI: 10.4192/1577-8517-v22_6

Sparer, L. (2024). *Token distribution in reward schemes.* BCCA, 91–98. DOI: 10.1109/BCCA62388.2024.10844480

Suan, D. L.Darwin L. Suan. (2024). Token approach in learning motivation. *EPRA IJ Environmental Economics, 11*(4), 45–56. DOI: 10.36713/epra17411

Sunyaev, A., Kannengießer, N., Beck, R., Treiblmaier, H., Lacity, M., Kranz, J., Fridgen, G., Spankowski, U., & Luckow, A. (2021). Token economy. *Business & Information Systems Engineering, 63*(4), 457–478. DOI: 10.1007/s12599-021-00684-1

Tan, L. (2018). Meaningful gamification & motivation. *Online Learning : the Official Journal of the Online Learning Consortium, 22*(2). Advance online publication. DOI: 10.24059/olj.v22i2.1167

Tartan, C. (2023). Blockchain welfare distribution model. *REGION, 10*(1). Advance online publication. DOI: 10.18335/region.v10i1.434

Vitai, Z., & Benke, M. (2021). Learning motivation in CEE students. *Marketing & Menedzsment, 55*(2), 45–59. DOI: 10.15170/MM.2021.55.02.03

KEY TERMS AND DEFINITIONS

Algorithmic Governance: Decisions handled automatically through rules baked into smart contracts.

Algorithmic Governance: when decisions get made automatically through rules baked into smart contracts, no central boss calling the shots.

Auditability: The ability to double-check that data or actions logged in the system are legit and to trace their full history.

BitDegree: A blockchain-based learning platform that started in Lithuania.

Blockchain: (covers permissioned blockchain, Ethereum blockchain): A shared digital ledger that's secure, can't be altered, and runs without a central boss, perfect for permanently storing data.

Cross-Chain Validation: Verifying a transaction that jumps between different blockchains.

Crowding-Out Effect: When too many outside rewards start killing the inner drive to do something just because it's cool or meaningful.

Curator: Someone who checks and green-lights contributions before they get locked into the system.

DAO: (Decentralized Autonomous Organization): A way to run things collectively with no top-down authority, using automated rules and open voting.

Dashboard: A visual hub showing contribution tracking and token history.

Digital Divide: Unequal access to tech or the skills to use it properly.

Distributed Governance: Decision-making and checks spread across a bunch of players instead of one central power.

DLT: (Distributed Ledger Technology): Tech that lets multiple parties share and verify a ledger without a middleman.

ERC-721: The Ethereum standard for minting NFTs.

Gamification: Weaving game-like elements into education, but focused on meaning, growth, and teamwork rather than cut-throat rivalry.

Gas Fees: The costs you pay to process transactions on a blockchain.

GDPR: The EU's rules on protecting personal data.

Gitcoin: A platform where the community chips in to fund open-source projects that actually matter.

GTC Token: The token tied to Gitcoin, used for voting and supporting those projects.

Immutability: Once data hits the blockchain, there's no editing or deleting it.

Interoperability: Different systems playing nice and swapping data without drama.

Layer 1 / Layer 2: The main blockchain (L1) and a secondary layer (L2) that speeds things up and cuts costs.

Learn-to-Earn: A system where you pick up skills or knowledge and actually get rewarded with tokens for it.

Matthew Effect: That snowball thing where the rich (in fame or resources) get richer, and the unknowns stay stuck in the shadows.

Micro-Credentials: Verifiable digital badges proving specific skills.

Micro-Deliverables: Those crucial in-between efforts that usually fly under the radar, like mentoring, coordinating, proofreading, or follow-ups.

NFT - Non-Fungible Tokens: (includes SBT - SoulBound Token): Unique, non-swappable tokens for certifying authenticity, identity, or skills. SBTs stick to one person and can't be passed on.

Node: A network participant who validates and records transactions.

Open Source: Code or resources anyone can grab, tweak, and build on together.

Pilot Prototype: A small-scale test version before rolling it out big.

Pseudonymization: Swapping personal info for IDs to keep identities safe.

Reputation Bias: The tendency to give more credit and attention to folks who already have a solid rep, even if others did similar work.

ResearchCoin (RSC): The token you earn on ResearchHub when your contributions get verified and approved.

ResearchHub: A platform where scientists share work and get it checked by peers, kind of like an open lab for ideas.

Research-to-Earn: Same vibe, but for science: you contribute research, it gets checked, and boom, tokens in your wallet.

Scalability: How well a system holds up when things get massive.

Smart Contract: Self-running code that triggers actions once conditions are met.

Solidity: The programming language for writing smart contracts on Ethereum.

Token: (covers fungible token, learning token, contribution token, recognition token, engagement token, GTC Token): A digital unit that stands for a contribution, a right, progress, or proof of something accomplished on the blockchain.

Tokenization: Turning an action or contribution into a traceable digital token on the blockchain.

Token-Weighted Voting: Votes where your say depends on how many or what kind of tokens you hold.

Zero-Knowledge Proof: A way to prove something is true without spilling the actual details.

Chapter 6
Operational Research, Transport, and Logistics in the Healthcare Sector

Wasswa Shafik
https://orcid.org/0000-0002-9320-3186
Dig Connectivity Research Laboratory (DCRLab), Uganda

ABSTRACT

This study presents an exploration of operational research, transport, and logistics in the healthcare sector, highlighting their critical role in optimizing resource allocation, improving service delivery, and enhancing patient outcomes. Efficient healthcare logistics ensure the timely distribution of medical supplies, pharmaceuticals, and life-saving equipment, particularly in emergency response and remote areas. Operational research techniques, including mathematical modeling, simulation, and data analytics, help streamline supply chains, reduce costs, and minimize wastage. Additionally, advancements in technology, such as artificial intelligence (AI), real-time tracking, and predictive analytics, are transforming healthcare logistics by enhancing efficiency and responsiveness. Challenges such as infrastructure limitations, regulatory barriers, and disruptions due to pandemics or natural disasters continue to hinder seamless healthcare logistics. This study examines innovative strategies and best practices in operational research and transport management for sustainability.

1 INTRODUCTION

The term "Operational Research" covers a wide range of scientific and analytical techniques and tools that are used to assist operational decision-makers in organiza-

DOI: 10.4018/979-8-3373-2975-8.ch006

tions from both the public and private sectors in making better decisions. Some of the methodologies used in operational research include modeling and optimization, forecasting, simulation techniques, statistical analysis, and many others. At the center of operational research is the use of mathematical models that can represent how a real-life situation could behave or should behave (Sordan et al., 2023). Based on these mathematical models, several quantitative techniques can provide insights into the real-life situation that is modeled so that better decisions can be made. The need for better-informed decisions means that operational research can be applied to the decision-making process of any organization, in any sector, and for any decision problem. One particular sector that can benefit significantly from the use of operational research techniques in healthcare (Singh & Parida, 2022). The need for better and more cost-efficient planning and allocation of healthcare resources has seen a significant increase in recent years, mainly due to two main reasons: the aging of the world population and the increased appetite for healthier living. The allocation of scarce healthcare resources to spatially distributed heterogeneous regions is necessary to ensure a good provision of health services to all citizens, independent of the geographical area in which they reside. These decisions can include the location of new hospitals, the allocation of patients to available healthcare facilities, the allocation of specialized medical disciplines to hospitals, and the routing of medical units (Marsilio et al., 2022). The terms operational research (OR), management science (MS), and decision science (DS) are often used interchangeably with the field of applied mathematics. This application is concerned with the improvement of operational systems, except that MS tends to be more concerned with business applications. OR is the title most commonly given to academic interest. It also has the broadest and neatest mathematical concepts, allowing the most coherent course structure for formal teaching (Alolayyan & Alyahya, 2023). DS has the broadest concept, offering better coverage of important unresolved problems; it is the title increasingly used by national and international societies. Both MS and OR, however, have the advantage of being recognized as broad fields of undergraduate university education, catering to wide career opportunities. Examples of possible applications include queueing theory for traffic and service problems, stochastic networks for various dynamical problems, and supply chain issues such as production, warehousing, and distribution (Jimmy Triputra et al., 2023).

Other important issues, such as public finance, advertising, and project management, might be considered under MS. The motivation of this paper is to document the widening applications in healthcare and its further implications. Successful information technology, fast telecommunications, and modern methods of business operations and decision-making based on management aim to pare down or cut down the outgoings—or to provide better service. All these are giving rise to a large shift in operational problems with significant interpretative and mathematical excitement (Cannavale et al., 2022). Distinct from traditional university research, which is usually pure, academic, and curiosity-driven and often has a long lag in follow-on applications, these new applied research activities have far-reaching consequences. Few practical companies can afford to take the risk of employing scientists who maintain the usual academic luxury of working on problems of no current consequences and are satisfied that the funding source is kept happy indefinitely by offering only long-term access to the final results. This is intended to provide an OR panorama to those considering applying some OR methods to the healthcare sector (Hussain et al., 2023).

Several authors have used mathematical models to solve problems in the healthcare sector, such as location models and assignment models. Since then, mathematical models in the healthcare sector have been proposed frequently. An optimization model is used to decide the admissions and transfers of patients for several cancer centers (the model aims to minimize the number of offered beds and trips with scarce or interrupted resources) and also the scheduling of patients being treated in radiotherapy (Dzando et al., 2022; Shafik, 2024o). The company behind this project is a company specialized in the development and application of software in management and decision support systems, enabling the coordination of activities in health institutions. An estimation of the number of ambulances required to serve a geographic area is made, considering the locations of historical clusters of traffic accident incidents. The optimization model used in the reconciliation of capacity and flow of demand in radiotherapy services considers priorities emanating from the medical literature and standard practice in healthcare, with a discussion on the implications and trade-offs of the introduction of schemes to guarantee access and/or delays incurred (Crowe & Utley, 2022; Shafik, 2024e). An optimization model for scheduling surgeries aims to maximize the success rate of surgeries and the overall utilization of surgical resources. Both models used algorithms and tools available in commercial solvers to evaluate the mathematical models. The demand for outpatient clinics in the healthcare sector is estimated in an academic hospital, according to individual characteristics and proximity to the clinic provision services, focusing on the absence of the patient. The model includes characteristics of the patients referred to each service and parameters to evaluate strategies for reducing the problems of incongruent demand-supply between clinics (Balasubramanian et al., 2023; Shafik, 2024h). An optimization model for multi-skilled staff at a hospital is developed, inserting variables in model constraints to allow the incorporation of management policies from the decision-makers.

A robust optimization model is presented to determine the number of cluster providers to meet the surgical requirements of a hospital, given the number of available surgeons in each cluster. The model is also used, but considering the distribution of surgeons in different periods of the day. An artificial neural network model is used to forecast the demand for hemodialysis treatments per day of the week with an approach for multiple outputs, updating the generated forecast over time with the real data (Dodge et al., 2023). The generated forecast is combined with an integer prioritization model to derive a monthly schedule, incorporating social welfare in the definition of the objective function to minimize patients' waiting time. An intelligent decision support system is developed to produce a proposal for physicians' duty rosters that satisfy all constraints in healthcare institutions. An optimization model is used to calculate a staff schedule for interdisciplinary physical rehabilitation teams (McDermott et al., 2021; Shafik, 2024d).

2 TRANSPORT AND LOGISTICS IN HEALTHCARE

Various companies in the healthcare industry conduct operations research, which is applied to transportation and logistics problems. Telecommunications and data management companies sponsor such studies and also supply solutions to healthcare institutions. Hospitals, clinics, social services, medical supplies, and physical rehabilitation institutions then enhance their overall performance. University-based teams conduct these studies using complementary competencies and resources (Leone et al., 2021; Shafk, 2023). Together, the solution providers, on the one hand, and the decision-making healthcare agents, on the other, represent a value chain fully oriented towards society. The optimization of healthcare supply chains contributes to, ultimately, saving human lives under stringent constraints in terms of resource costs and availability, as well as ethical and legal rules. The logistics and transport within a healthcare organization must deploy reliable and trustworthy solutions in terms of performance excellence and social responsibility. Specific features of the healthcare sector, including the idiosyncrasies of its physical and human activities, demand special attention in the configurations of supply chains (Shafik, 2024k; Tippong et al., 2022). The companies offering those logistics and transport services must also participate in partnerships in which operations research contributes to the overall objective set by healthcare organizations. Councils regulate entry conditions. State-owned enterprises and non-profit organizations deliver reliable, trustworthy, and universal services, even if consumers are not willing to pay for the often-unperceived value. For-profit enterprises must deliver universal, reliable, trustworthy healthcare services and not only consumer surplus that varies significantly with their willingness to pay (Diwas Singh et al., 2020).

2.1 Importance and Challenges

Healthcare logistics is becoming an increasingly important subject within operational research. This is driven by various factors, including the public and political need to ensure the efficient operation of healthcare systems, the sheer size of many healthcare systems, the relatively low level of performance in terms of access to healthcare, the effectiveness of the clinical services offered, and the overall degree of patient satisfaction. Furthermore, while the area of healthcare logistics is by no means a recent addition to the operational research portfolio, many of the innovative logistical research solutions are (Jha et al., 2022). This is not just down to pioneering research, which is undoubtedly crucial. The increased application of established operational research techniques, as well as supply chain ideas from other industries and services, is just as vital. In many cases, these can offer significant and beneficial contributions to healthcare logistics. The obvious challenges in healthcare logistics include issues of patient access to healthcare services, treatment of patients during the care process, reverse logistics, and discharge. This last category of healthcare logistics is a broad one, including a range from patient transport and network planning to pharmacy distribution (Lamé et al., 2023). All of these areas are clearly crucial since the overall goal of the elective infrastructure is the speedy and effective treatment of patients in an organizationally efficient manner.

2.2 Key Concepts and Techniques

This refers to the concepts and techniques of operational research that have been widely used in healthcare logistics. These concepts help in understanding solutions to many problems and the development of proper models that may lead to better offers to patients, whose interest is a main driver in healthcare logistics. The logistic decisions in healthcare networks have been mainly presented here, such as the location of healthcare facilities, districting, assignment of facilities, and resource allocation (Keskinocak & Savva, 2020). Nonetheless, operational research also offers solutions for problems such as the appointment system, bed management, patient flow, and emergency healthcare. Optimization, queueing, and simulation are the main techniques that have been used. Drawing the main focus on optimization, we used the following techniques: location science, mainly considering the continuous and discrete location problems; network science, mainly applied to healthcare network problems but also to the cupola problems and communication problems; combinatorial optimization started considering the e-hub location problem; integer programming was also applied; heuristics, or simply models in which the results were used when the exact solution was not possible (Shafik, 2024c). Variables and the objective function were used to model the problem, and the constraints expressed relationships existing in the problem. There were three main 'families' of constraint types:

(i) constraints existing because of demand;
(ii) constraints existing because of space;
(iii) constraints existing for the operations.

3 INTEGRATION OF OPERATIONAL RESEARCH AND TRANSPORT LOGISTICS IN HEALTHCARE

In facing a global challenge such as managing and mitigating the effects of a pandemic, better understanding the role hidden from the human view of health supply chains offers a critical lesson for designing a more effectual and resilient healthcare system. For example, the pandemic has greatly accelerated telemedicine and the use of technology, and it has also led to an unprecedented operation of health supply. It is true that without logistics, healthcare is zero (Shafik, 2024n). Likewise, the pandemic experience has helped demonstrate the principal role that applied operational research and technology can bring to optimizing strategic decisions and, directly affecting emergency needs and effectively treating a considerable number of patients while at the same time contributing to the development of new drugs, testing kits, quarantine facilities, and vaccines. The main contribution of the present paper consists of providing insights into a health supply chain model constituted by patients, healthcare specialists, warehouses, and equipment, all governed by transport logistics and operational research guidelines (Nair et al., 2023). We propose that healthcare deliveries must be deeply considered logistics services providing maximum safety, speed of response, regularity, and appropriateness of the medical goods delivered. In a broader sense, future research will be processed by expanding our model to include other paradigms and by adding other demand nodes, as well as their interactions. If health supply optimization could be ensured, healthcare would benefit from belonging to an optimized supply chain. As we have seen from the current pandemic, undoubtedly, healthcare is not zero, and logistics may save lives. Moreover, effective prevention strategies proven to diminish risks from cardiovascular diseases, obesity, lung and breast cancer, anxiety, depression, and other related pathologies should be promoted. Applied operational research plays a principal role in affecting strategic decisions, solving the most field-oriented problems, implementing and achieving specific aims, and, at the same time, contributing to the development of new drugs, testing kits, quarantine facilities, and vaccines (Dodge et al., 2023).

3.1 Benefits and Opportunities

Operational research, transport, and logistics methods provide decision support in areas where the optimization involves a large number of transactions that must be scheduled, routed, and worked. This includes many processes, such as planning and scheduling, that occur regularly and involve many stakeholders across the EPR. The cost of such support can be high, but the cost of mistakes or poor planning can be massive—in terms of wasted resources, lost time, and, mainly terms, suboptimal outcomes for patients. This looks at some currently successful applications and then moves on to present more evidence of missed opportunities. The purpose is to show that operational research and related methods can offer much more to the health service. This demonstrates numerically that the systematic application of operational research methods within health service delivery could contribute to significant system efficiencies, savings in costs, and benefits for patients (Kakalou et al., 2018). Furthermore, the demonstrate the parsimonious nature of these methods in order to encourage interest and future applications within the wider healthcare community under the phrase, no budget? But need results? Try operational research. With such a background, in the sections to follow, the methods and models applied to the health service sector are summarized, and some of the results of each of these are presented, analyzed, and discussed. We discuss some potential areas of future research and development (Bhat et al., 2023; Shafik, 2025b). Throughout the chapter, alongside method form and function, this expositional approach takes into account the computational processes that led to the model outputs and decision support—so as not only to provide a summary of the results but also to illustrate the related economics. In this way, we also contribute to the role of the operational researcher as a presenter.

3.2 Case Studies

A study initiated by a health authority used optimization as the driver of radical transformational change in the clinical services of Edinburgh's six acute hospitals. The overall aim was to develop service provision, identify more cost-effective processes that minimized the number of operations, and increase successful patient discharges and overall patient well-being. Two operations were successfully developed. The first was the introduction of a new clinical specialty, called colorectal nurse practitioners, which focused on the non-clinical care for non-complex surgery patients, allowing them to be discharged on the day of their operation, thus doubling the capacity of the hospitals for this surgery (Carter & Busby, 2023). The second was the introduction of the cardiology day case model, which enabled additional patients to use this service without the need for additional resources. In 1995, a cost-driven network model for planning cataract surgery was developed. The demand for surgery and the constraints given by health resources were included to ensure that the highest number of cataract interventions was carried out within a one-year schedule (Keskinocak & Savva, 2020). The situation in the Netherlands indicated that a large capitation system was in danger of producing a high volume of cataract surgery. The idea of the model was to investigate the capacity of surgery flows on a regional basis in order to show the impact of this kind of intervention, which should be used in the political administration of a hospital or of a health authority to optimize available resources without increasing costs. The specialized scientific support and solutions that the study provided for other hospitals showed what quantification applied to allocation policies could provide in terms of efficiency and effectiveness of the interventions provided by the public health service (Fusco et al., 2023).

4 OPTIMIZATION MODELS FOR HEALTHCARE LOGISTICS

We first provide a general review of operations research in the healthcare sector, highlighting the main contributions in terms of problem modeling and solution methods used. A brief comparison of what is currently being done in each of the areas in two economic realities is used to emphasize the potential bigger impact of proposed research in other countries. We then cover in detail six specific problems of interest with more detailed information in three main dimensions characterized by the literature on optimization and the application to healthcare logistics (Walsh et al., 2018). These focus on bed management, team care-based schedules, surgical waiting lists, patient flow, telemedicine, and medical waste management. We finalize the paper by addressing potential application areas and research perspectives. The demand for improving the public sector of healthcare and reducing its budget has led to cost optimization is one of the main challenges in healthcare over the last few years. Operations research has an important role to play in defining and achieving those important goals in an area with clearly defined features characteristic of service operations. The main contribution of operations research to healthcare is not expected to be on the definition of new methods but on the application of existing ones and adjustments of the modeling that will best address the constraints and, at the same time, provide a better boundary on the global optimum (Pradhan et al., 2021).

4.1 Types of Optimization Models

The objective of this section is to present an overview of operations research modeling and optimization techniques and methodologies used to design and analyze healthcare systems and related logistics and transportation problems. These are wide-ranging and management-intensive problems. Most tasks fall within the realm of facility location, distribution network design, equipment and vehicle planning, scheduling, and route optimization. In particular, most of these problems can be formulated as mixed-integer linear programming models (Detwal et al., 2024). Such a formulation ensures that both the allocation and attributes of facilities, as well as the routing and scheduling over a time horizon, are determined with the objective of minimizing a certain cost function, subject to constraints. These models are the perfect tool for decision support systems exploiting the functionalities of commercial solvers that guarantee satisfactory performance in terms of solution stability and computational speed. Moreover, the effects of transplants and harvesting coordination, inventory costs, parcelization, and routing are often required. Nonlinear problems include demand forecasting and patient-arrival processes, whereas stochastic models solve issues when demand and supply conditions are uncertain (Saraswat et al., 2022). In order to illustrate the applicability of operations research techniques in healthcare, thorough insights into the problem types characterizing the healthcare setting are presented, and future model types are enumerated.

4.2 Applications in Healthcare Settings

Given the increasing interest in OR/TS applications to healthcare issues, the following sections address two key factors. Section 4.2 reviews works that apply OR/TS methodologies to healthcare settings other than logistics and service improvement, while Section 4.3 focuses on logistics and service improvement in healthcare settings. We review applications of OR/TS methodologies in a variety of healthcare settings (Leite et al., 2021). We will include under this heading papers that focus on healthcare processes in hospitals, including the use of operating rooms, beds, and other hospital resources. While most of these applications are conducted in large hospitals, others focus on health centers, care homes, or any institution where patients must stay for a while to receive care. We will also include here some papers that focus on logistics in other areas, such as the coordination of donations to increase donations, recent work to operationalize personalized medicine, and the location of defibrillators (Mahjoob et al., 2023; Shafik, 2024l).

5 DECISION SUPPORT SYSTEMS IN HEALTHCARE LOGISTICS

A Decision Support System (DSS) is a human or computer system that is used to support the decision-making process. The DSS includes many different types of models that allow the user to answer the "What if X?" questions and interpret the results of the model. The decision-making process involves problem structuring (defining the problem to be solved) and problem-solving (using different models to solve the problem). The coordination of both processes is more efficient with the use of a real-time DSS since the user is able to analyze and test the modifications of the model in progress (Minh Dang et al., 2019). Solving a problem implies the definition of management strategies (i.e., the base for the operational plan). The comparisons of the results obtained and those obtained in this chapter will show the usefulness of DSS in healthcare logistics. This chapter will be based on the methods outlined above in order to present decision support systems in healthcare logistics. There are many different types of decision support systems available that can be used to support logistics decision-making. In the following, a DSS in logistics for the healthcare sector will be presented along with a case study validating the efficacy of the DSS within the service sector (Ramalingam et al., 2024; Shafik, 2025c). The implications for small to medium-sized providers focused on hospital logistics will also be presented.

5.1 Definition and Functionality

Following from the initial half of Definition 1.1, which characterizes applied OR in general, and in the last subsections, which examine its topic differently from other approaches, the focus of this subsection is on the first part of the involved definition. Specifically, it is concerned with the operational research field and, more precisely, with operational research, transport, and logistics, particularly when it is carried out in the healthcare sector. First, however, some more background is necessary. The phrase 'operational research' is in widespread use throughout the world. Constraints of language and culture, however, result in several idiosyncratic terms being used (Syahrir et al., 2015). Not all refer accurately to our vast discipline. We now provide a more comprehensive definition of operational research and then suggest its scope by enumerating eleven areas of application.

5.2 Implementation and Benefits

Healthcare systems all over the world are facing severe challenges in the delivery of care to people. These challenges include rapid medical innovations, increased regulation, new payment models, and the complex nature of organizational arrangements. Largely because research in operational research, transport, and logistics has been useful in other sectors, the application of operational research in healthcare has been of growing interest recently. What, therefore, makes operational research not useful in the healthcare system (Sachdeva et al., 2007)? These could include an imbalance in the development and application of methods, challenges in the quality of data, the role of professional organizations and societal factors, and a lack of preparedness to train healthcare professionals in health operations management. Healthcare is an industry with unique challenges and diverse entities. While some of these entities are in the public sector, others could be under private management or are a partnership between the public and private sectors. Public and private healthcare providers both compete against and complement each other (Crowe & Utley, 2022). Similarly, public and private payments are made to healthcare providers. Public and public-private partnerships can promote or manage healthcare brands related to the image of or support from the community. The following are equally diverse: purchases of health care in the private sector, buy and sub-buy from the public and private sectors, and insurance. Of this diversity, differences in healthcare arrangements among countries introduce additional complexity, and these can even cause differences within countries based on economic status, culture, and policies. Nevertheless, while the challenges of health operational management can be diverse, the meaning is a common denominator for health operations in general (Mahjoob et al., 2023). This offers examples of the solutions achievable through the application of operational research in solving these challenges.

6 QUALITY IMPROVEMENT IN HEALTHCARE LOGISTICS

This article provides an overview of quality improvement using operational research, transport, and logistics in the healthcare sector. It summarizes how these techniques can be used in different sectors and, in each sector, provides certain examples of research in the areas and the corresponding operations, transport, and logistics problems. It concludes with a brief discussion about the role of operational research in quality improvement in the healthcare sector and future areas of research (Detwal et al., 2024). Quality improvement is complementary to quality assurance. The former has a focus on error reduction and variance reduction. Quality improvement techniques include the use of data to understand performance measurement, information that is needed to diagnose a problem, and a system to support improvement. Additionally, quality improvement provides an organization with a systematic approach to problem-solving and helps an organization learn from its mistakes and prevent recurrences (Patel et al., 2024). Quality improvement is a systematic approach that utilizes study to improve an organization or delivery process or decrease the number of problems once the diagnosis is made.

6.1 Concepts and Tools

Health is a central condition for people's well-being. The improvement of the health system is an engaging concern of the governments, which have sought to offer the population a faster and more efficient public health service, particularly for those sectors that are economically more productive. The models used for the study and improvement of health systems were developed within the field of Health Service Research, a disciplinary area traditionally characterized by an interest in economic, social, demographic, or nursing aspects of patient outcomes (Kraus et al., 2021). In this paper, we present, within the framework of Operational Research, the models and techniques that are developed and/or can be profitably applied to support and solve the decision-making problems that exist in the healthcare sector. These models and techniques are often generalized and tested in different application contexts, both in health and in other sectors. In particular, we illustrate several OR models and techniques, focusing on drawing a picture of the most developed ones for the health field. First, we refer to the models and techniques traditionally used to solve these planning problems; then, we give an overview of the specific problems of logistics and intermodal transportation that exist in health services (Hallam & Contreras, 2018).

6.2 Measuring and Monitoring Quality

Many healthcare systems conduct regular large-scale surveys of patient satisfaction or patient experiences, both at organization and aggregate levels. A large number of organizations also regularly monitor the outcomes of their medical procedures and develop periodic public reports on their performance. Data from these surveys provide information about areas in which hospital management excels and areas for potential improvement (Rais & Vianaa, 2011). The interest in such measures is being driven by the increasing demand for patient information and the concomitant interest in the purchasing power of strategic purchasers. Indeed, the provision of quality care is mandatory, and their performance status is published. While national government institutions look to standardized league tables, benchmarking is often performed at a more local level. For instance, managing within a region or hierarchically organized services such as those for cancer have requested evaluations not only with a provider's peers but also at the specialist cancer service level (Ur Rasool et al., 2023). It is argued that performance indicators only make sense if they are useful for improving patients' outcomes. Although organizations should not ignore indicators that they are falling behind their peers, one trusts that the examination and potential improvement of actual processes will dominate management efforts. Performance measures of service quality are not exclusive to the UK; reports of several projects have cast a wider net. Results from these studies suggest that the rigid use of any performance measure can lead to perverse situations. For example, encouraging early discharge of patients or converting some inpatient treatments to day cases (Keskinocak & Savva, 2020).

7 RISK MANAGEMENT IN HEALTHCARE LOGISTICS

Risk management has several definitions but can be considered as the identification, assessment, and prioritization of risks and the coordinated application of resources to mitigate, minimize, and control the probability or impact of these threats or exploit the opportunities in healthcare logistics. Risk management can also encompass the optimization of risk-reward intersections. Health sector logistics has certain problems; attainment of the least time and least cost factors, because of the short time available for providing life-saving services, is deemed to be influenced by risk factors (Ramalingam et al., 2024). Prevention of risks is preferred to the management of processes created by them. Suppose the logistics process can be planned in such a manner as to render the negative influence of risk factors low rather than managing the risk. In that case, the reliability of systems can be maximized, and better operations can be performed. Activities with a high risk of inaccuracy must be re-evaluated, and the most potentially irrational processes should be removed completely or, at least, optimized. In the framework of a general operational research approach, whatever the output of our logistic analysis is, resource costs can be used as one of the integrative components, and zero-cost risks are also incorporated properly by energetic business management stakeholders (Sordan et al., 2023).

7.1 Identification and Assessment

Aiming to present the current state of development within the increasingly important and rapidly growing field of healthcare logistics and lean production in the healthcare industry, with special emphasis on the operational perspective. Our database of scientific studies, consisting of 33 publications, allows us to define the main areas that should be addressed by the operational research community in the domain of healthcare logistics, as well as the gaps and key challenges that remain open. As a result, a wide range of opportunities can be identified for future research, including a number of concrete research questions that can be the basis of future research in healthcare logistics (Carter & Busby, 2023). The operational research community identifies research topics, research questions, and methodologies that seem to be better addressed according to the available state-of-the-art techniques. By exploring these aspects, researchers may help to fill critical gaps in the literature and develop operational models and algorithms that are able to provide better results. This interesting research domain can give a new perspective to the advances and highlight the connections between surgery and healthcare processes and the more general field of logistics (Rais & Vianaa, 2011).

7.2 Strategies and Best Practices

There is little to draw on in the literature to support a discussion of best practices in the application of operational research methods to transport and logistics in the area of healthcare. However, consultation with practitioners and analysts who have first-hand experience provides a range of experiences from which best practices are extracted. Drawing on this exercise, we identify some widely agreed best practices in the application of operational research to these issues in the healthcare sector in three categories: consultancy commissioning, model building, and knowledge transfer (Sachdeva et al., 2007). It is widely agreed that it is essential for a flexible approach to consultancy arrangements. Because of the wide gap in skills and expertise between us as academics and the healthcare sector, there is a common need for rapidly developing work that starts from both a strict information base and a focused application. Consequently, we also need to be ready to begin the work, and if it should quickly accelerate, where some rapid analysis has uncovered additional practical information and knowledge. This includes the experience and practicality of the systems that derive from knowledge management (Lamé et al., 2023). Illustrative consultancy models that others draw on include teamwork and continuing professional practice.

8 SUSTAINABILITY IN HEALTHCARE LOGISTICS

Sustainable healthcare logistics and reverse logistics are covered. These include environmental and social aspects and discuss these in the context of decision support, considering criteria used in decision support systems, including environmental, social, and economic measures, in a wider conceptualization of logistics. Future healthcare logistics services should be made more sustainable. However, extant decision support systems and optimization models in healthcare logistics are based on single and non-integrated strategies supporting predefined logistics functions regarding patient care and health outcomes (Lamé et al., 2022). They usually disregard the potential synergy among these logistics strategies and neglect to consider significant dimensions of sustainability. Previous studies have identified a number of logistics processes implemented in hospitals that might facilitate hospital logistics performance, such as enhancing the movement of goods to the patient service/care units, reducing transportation costs, and releasing time for staff to concentrate on direct patient contact. There is potential for enhanced performance as these studies have not integrated these processes within the wider logistics operation, taking into account their potential role in enhancing economic, environmental, and social dimensions of sustainability (Ilangakoon et al., 2022; Shafik, 2025c).

8.1 Environmental Impact

Logistics has been acknowledged to have a significant contribution to environmental problems by generating externalities and negative impacts on stakeholders. The external cost of transport is a key factor that must be taken into account in transportation operations management. The variety of measures that affect the environmental and social costs of prolonged transportation in healthcare logistics is analyzed, studied, and assessed. As a result, an operational decision-making tool is proposed, and its main features are detailed for addressing the potential of intermodality, load factor, and consolidation of flows to control and reduce such impacts in healthcare, thus achieving sustainable development targets (Mahjoob et al., 2023; Shafik, 2025a). Healthcare logistics accounts for a substantial portion of international merchandise trade, particularly in low- and middle-income countries. The sector is known to be conservative and an expensive part of the economy, showing lower productivity and efficiency gains. The reduction of transport-related economic externalities is considered a priority, given the high amount of human and social capital involved. Public healthcare sustainability, together with achieving high standards of accessibility, affordability, and quality from the public perspective, is one of the most critical public tasks (Shafik, 2024a). These considerations, along with the explosion of global health and the rise of the world population, came as significant drivers for the search for new and innovative technological solutions to improve efficiency and lessen the environmental costs of healthcare logistics on every related transportation chain network. The proliferation of new innovative technologies provides a multitude of possible logistics solutions that may yield substantially improved performance, together with the lowering of operations and control costs borne by stakeholders. However, the implementation of such innovations in healthcare logistics, particularly in last-mile distribution, still presents significant hurdles and imposes structural changes in public and private service organizations, related infrastructure, local governments and authorities, consortia, and contracts (Shafik, 2024b). All mentioned stakeholders require change or incentives to move towards efficient, environmentally friendly, and sustainable healthcare transportation solutions in the contemporary world.

8.2 Green Practices and Initiatives

Several green practices have been fostered or initiated in the healthcare sector. Among them are recommendations, the substitution of hazardous chemicals with safer alternatives, as well as healthcare accreditation initiatives. Accreditation means an approach to the assessment and upgrade of the quality of healthcare facilities and staff, and a certification of their quality. In the United States and Europe, the federal government is the major healthcare provider, and it has already established targets for implementing cleaner practices within its facilities (Shafik, 2024i). These green practices are typically target-specific and focus the responsibility of environmental control and pollution prevention on healthcare service providers. However, only more specific and rigorous regulations are likely to provide overall workable disciplining mechanisms for healthcare logistics 'polluters,' such as pharmaceutical and medical agent producers and shippers, providing a more level playing field and identifying accountability and responsibility in the field. Accidents and emergency planning could also be used as a tool to promote better healthcare logistics practices (Shafik, 2024f). The cleaner practices and recommendations implemented in different healthcare facilities around the world to help preserve the environment have produced significant savings, which have pointed in the green direction that it is possible to both safeguard patients' health as well as the environment at lower overall costs. At the micro level, several healthcare facilities started to realize that better pharmaceutical agents' inventory management, end-of-life accounts management, and more sophisticated logistics provide reasonable cost savings, which are certainly measurable. Indeed, to acknowledge these more successful experiences, a program was established that rewards hospitals that take initiatives to reduce their environmental impact (Patel et al., 2024). Moreover, other innovative healthcare logistics initiatives, such as the only drug supplier to hospitals in Portugal, now using containers for drug shipments made of biodegradable and dye-free vegetable materials, are now starting to benefit from growing new market opportunities, given that better environmental performance is now seen as a competitive advantage. Costs certainly play a major role in the institutional decisions of healthcare logistics, but the inclusion of indicators reflecting societal performance in health, environment, and safety could also help underpin some value from that performance perspective (Fusco et al., 2023).

9 TECHNOLOGICAL INNOVATIONS IN HEALTHCARE LOGISTICS

Our proposed technological innovations in healthcare logistics embrace smart contracting technology between hospitals and their suppliers, the use of smart logistics robots for handling diverse logistics operations, and real-time data management tools to gather information from different healthcare parties and ensure the best decision-making and performance. It is widely argued that smart contracting technology can improve working partnerships and coordination between organizations, reduce service failures, address emergent or unstable unplanned supply chain scenarios, and support effective change and innovative development (Hussain et al.,

2023). Both smart contracting and smart logistics robots are emerging and rapidly changing technologies that could be utilized in the healthcare logistics environment to support performance in an integrated and coordinated healthcare sector. The use of both technologies is a niche within the healthcare sector, and further research is needed to define, refine, and advance our understanding to enable a smooth take-up, development, and implementation of their full potential within different healthcare supply chains (Balasubramanian et al., 2023).

9.1 Emerging Technologies

This details the latest advances in the use of operations research and logistics within the healthcare sector. The systematic approaches have been applied to new areas that were not feasible just five years ago. There are many new initiatives led not just by academics but also by health professionals and other commercial stakeholders. The wealth of new data sources within the healthcare sector enables much richer models to be applied to both traditional problems that are finally being solved well and new opportunities that were not previously possible (McDermott et al., 2021). These include new techniques for handling queueing at hospital clinics, surgery optimization, stent routing within hospital networks, drug development, and ward routing of products. The chapter concludes by detailing the new areas of healthcare that we are on the cusp of solving as challenges. Key directions from recent initiatives include logistics focused not just on cost reduction but also on runtime. A systematic study of factors that explain why the increasingly efficient healthcare sector appears to be inefficient has led us to wonder what is the equivalent of Moore's Law in the world of healthcare? While major technology changes will bring many opportunities in terms of future innovation, data handling, and logistics, these are happening now; systems are being deployed and are making healthcare more effective and more efficient. For example, the very rapid diffusion of mobile health technologies, such as patient monitoring, has opened up the sector for logistic opportunities, such as logistics focused not just on cost reduction but also on runtime (Nair et al., 2023). Items are also being administered to patients; we are beginning to understand these items' interfaces so that we can apply these items or their underlying technologies to new types of problems. The notion of the 'item of logistics' has evolved from making the push-tray model of hospital logistics work efficiently within and between our hospitals to enable initiatives that extend beyond the boundaries of those hospitals. There is much inefficiency within the logistics interfaces that are being administered because we lack intermediaries that mediate complex problems: now is the time to research these mediators (Jha et al., 2022).

9.2 Applications in Healthcare

Operations research, transport, and logistics methods have been used in the health sector for some time. Initially, health applications were focused on the planning of medical facilities - for example, their location and their size (the number of beds, etc.) - through the use of location and capacity planning models. This reflected the way in which operations research and logistics were being used in other sectors of the economy and the close relationship with the discipline of economic geography that is relatively common to operations researchers. Since the mid-1980s, however, there has been a steady growth in the use of operations research, transport, and logistics methods at the interface of the provider and client sides of the health sector (Fitzpatrick et al., 2017). This is perhaps not surprising as the movement of patients and clinical staff (both within and between health care services), and the organization and delivery of services to enhance the health of the client who also happens to be a patient are areas of technical expertise within the field of operations research, transport, and logistics. As we discuss in various sections of this chapter, the active use of operations research, transport, and logistics methods in the health sector can be more readily and clearly demonstrated in some areas of the health sector than in others (Mahjoob et al., 2023). The use of operations research, transport, and logistics models has been quite narrowly focused geographically as well. To some extent, this is arguably an appropriate and responsible time and geographic focus: lessons that were learned and best practice models that were established in health care service delivery and management when practices and management styles were exported to other countries were found to be equally applicable to the different healthcare delivery environments in that they were applied with very little adjustment or adaptation to other, primarily European, contexts (Yu et al., 2021).

10 DATA ANALYTICS AND ARTIFICIAL INTELLIGENCE IN HEALTHCARE LOGISTICS

We begin with data analytics and artificial intelligence in healthcare logistics. The use of electronic records in healthcare over the last few years has provided an opportunity to mine a plethora of data in view of either improving the care of admitted patients or for various operational purposes. This chapter is focused on the latter part. Studies into ward rosters, theater schedules, and hospital bed inventory problems are considered in this chapter. Various solutions based on operations research and computational techniques, such as data mining, optimization, and queuing theory in the realm of care, are considered (Wiig et al., 2020). This chapter illustrates how problems in hospital logistics are modeled mathematically and are employed to utilize health informatics data available in hospitals. Additionally, the reader is also informed about the critical decision problems that can benefit hospital management and others. After understanding the patient admission data, this chapter also addresses some exploratory data analytics that can be done for hospital ward bed allocation, operating bookings, and others. Currently, there are very few hospitals that employ queue management, and the promotion is limited to larger hospitals. These were done in operations research. More often than not, a central booking office would fix the operational schedule of theaters. In the second part of the chapter, the reader is introduced to several different aspects of data visualization and exploratory data analysis forward and bed allocation (Pearson et al., 2013). Though ward bed allocation and operating bookings may be perceived as static, this is always threatened by the arrival of bed requests. Some dynamics of patient bed allocation operating theater scheduling, and emergency case scheduling are discussed.

10.1 Role of Data Analytics

Technical advances have made large databases available in healthcare. There is great potential to use big data and analytics in order to improve healthcare services and patient outcomes. Research in information systems has been investigating the use of big data for some years, notably in supply chain management and customer relationship management. The potential is great, encompassing screening of the healthy population, improving service delivery, enhancing the decision-making of health professionals, and supporting health service management (Palmer et al., 2018). Patients and citizens can also use big data for their health and healthcare. However, there are many challenges and constraints to realizing this potential. This includes technical issues such as improving interoperability, combining and analyzing diverse data types, and ensuring data quality and data privacy. This exciting and potentially transformational field must be thoroughly grounded in health and organizational theory (Kraus et al., 2021). Statistical methods such as linear and logistic regression analysis play an essential role in clinical trials, cohort studies, and randomized controlled trials. However, they have a limited role in settings with observational data where the set of predictive or independent variables closely reflects the practically available covariates or envisioned predictors (Diwas Singh et al., 2020). Ordinary matching-based and inverse probability-weighted estimation-based methods ensure confounding control and reduce bias by using the entire empirically estimated distributions of the confounders. These methods are increasingly available through data analytics tools that produce easily interpretable estimators while dealing effectively with unmeasured and undefined confounders. They can also enable the detection of differential treatment effects and a precise assessment of effect heterogeneity and questions related to external validity (Ilangakoon et al., 2022).

Healthcare services worldwide are faced with the challenges posed by the super-aged society, where older people (aged over 65) make up a substantial, ever-increasing portion of healthcare users. In Hong Kong, the percentage of older adults will rise to more than 30% by 2033. At the same time, the number of carers available to cater to them is decreasing significantly. Based on the above trend, to cope with the surging healthcare demand in the city, the need to cater to a large number of elderly citizens across a varied spectrum of service needs and diseases in efficient and effective ways is acute (Carter & Busby, 2023). Efficient healthcare requires comprehensive, continuous, and close monitoring of patients at the primary care level, including the regular management of chronic diseases. In response to such demands, big data research for healthcare, namely healthcare big data analytics, has been an important research domain in operations research. The technical work includes various statistics-based data prediction and forecasting techniques and operations research-based optimization techniques to address different spectrums of healthcare issues in ailment prevention and management, and elderly monitoring (Pearson et al., 2013).

10.2 AI Applications in Healthcare Operations

AI logistics is a successful and dominant field of research in artificial intelligence and operational research. Practitioners are pushing towards real-time decision support as a cornerstone of logistics automation. Health is a socially beneficial domain that has no spatial bounds and is the essence of a multitude of practice-based and urgent environments, two characteristics that are sorely lacking from the existing case studies. In this section, we showcase a few of the more than three decades of successful research in healthcare logistics supported by AI techniques, and we contrast this success with the lack of new work in emergency patient recovery and bed scheduling in real hospitals (Shafik, 2024m). We argue that there is no substitute for real application at the sharp end to push the field forward, but equally important is the sharing of current tools with current hospital information systems. Healthcare can be considered a complex adaptive system where interconnectedness across all services and the patient at heart dictates response-led local adaptations that must constantly manage and fine-tune resources. Users also demand 24/7 service provision and powers of emergency intervention and lifesaving. Although a lack of hospital beds is not a healthcare issue, the uncoordinated movement of mental and physical patients from one-bed type and hospital to another hospital bed and unit is a well-known bottleneck in the efficient emergency patient recovery process (Carter & Busby, 2023; Shafik, 2024j). An unfilled or uncoordinated path from admission to discharge offers nothing to patient flow. Closed-loop scheduling is also an unresolved, real-time issue.

11 HUMANITARIAN LOGISTICS IN HEALTHCARE

In recent years, there has been an increased amount of literature focused on humanitarian logistics to examine how personal or financial support can be provided to unfortunate people through the design and control of supply chains that can respond to the needs of the disadvantaged. Humanitarian activities are traditionally based on the careful review and use of methodologies. Thus, the connections made with respect to humanitarian applications from the areas of operations research, transport, and logistics lead to the definition of the new area of research known as humanitarian logistics (Shafik, 2024g, 2025d). The specific application of both traditional and recent transport systems to the field of disaster healthcare relief is shortly presented in this chapter by using real-case applications or disjoint examples, and the main decisions are grouped into those that can be concentrated on transport and those that can be focused on the purpose of humanitarian operations. The effectiveness assessment of humanitarian actions regarding distribution decision-making, cost management, and the social implications that have an impact on the life improvement of disadvantaged people due to the activities carried out by humanitarian organizations are research topics of great relevance and wide generalizability (Saraswat et al., 2022). In such a context, the relationship between humanitarian activities and operations research is very strong, as such methodologies are well suited to face organizational challenges and to provide concrete and effective tools to be completed in an appropriate environmental and social context.

11.1 Definition and Scope

As in the rest of the economy, the use of operational research in the healthcare sector is not new. In the 1950s, an application to a cancer center included methodology using the Delphi technique, queuing theory, Monte Carlo simulation, decision theory, and phasing to estimate the construction costs of the center, while presenting equations to minimize the personnel costs, inventory costs, and costs derived from the use of the linac. The applications of operational research multiplied with the growth in healthcare costs and the possibility of using more complex techniques (Crowe & Utley, 2022). In the last 50 years, we have seen a growing interest in using these techniques in the field of health and in the development of specialized journals. Nowadays, the founding of a section on Health Applications represents a success. These applications are, in a certain way, more specialized and complex versions of applications that were already being implemented in other areas such as demand management, inventory, asset management, or HR without specializing in the healthcare sector. The specialization in the healthcare sector is justified by unique characteristics of a business that, besides being extensive in terms of human and financial resources, is sensitive because it provokes affective reactions. Furthermore, it is also a challenging field because of specific characteristics very different from other traditional fields. These include, among others, the large diversity of services, the existence of a public-private duality, and the need to achieve competitive advantages (Shafik, 2024b; Syahrir et al., 2015). All these aspects justify that if approaches to operational management in the healthcare sector had incorporated the growing observations.

11.2 Challenges and Solutions

Several factors are contributing to increase the importance of Information Systems and Information Technologies in Healthcare Sector. Some of the most relevant reasons are:

(i) the growth of the number of suppliers and health providers in a process known as 'outsourcing';
(ii) the need for improving the service to the patient by introducing systems for Health Records;
(iii) the new requirements in terms of confidentiality and privacy in the medical history area;
(iv) the strict liability of the doctors when the processes are not managed according to some basic quality rules in a possible environment if the information is easily available;
(v) the need for using patients as a means to feed the research and teaching processes committed in the services. However, these researchers have to assure and warrant the integrity of the patients and their data in a near future if they want to continue this support;
(vi) the introduction of smart appliances for the control of critical and vulnerable groups, e.g. the elderly or individuals grieving from chronic diseases. From all these goals, the one that has been less tackled has been the one related to improve the service using the systems of Information, unfortunately, a large part of health professionals consider that the Information Systems is conceived with another intention: to increase the productivity of season personnel.

12 INTERDISCIPLINARY APPROACHES IN HEALTHCARE LOGISTICS

In a world of growing complexity and tight resources, medical professionals must concern themselves with increasing workloads and the pressures of performance culture. For this reason, there is a need to seek help from other non-medical fields, including business sciences and logistics, to focus particularly on the management of the resources made available to healthcare. In modern medicine, significant factors either hinder patient flow or prevent optimal patient care, having become frequent obstacles that result in unnecessary waste (Shafik, 2025e). Thus, the healthcare system continues to swell in size and strives to keep costs down in a highly competitive environment in order to satisfy patients and staff concurrently. Logistics, a business science that uses analytical tools and decision support strategies for the benefit of the operating system through the availability of necessary resources, infrastructure, and operations, plays a significant role as a tool to address challenges, working on quality improvement and cost reduction in health system management (Sordan et al., 2023).

The term "healthcare logistics management" refers to the objective of this science: to effectively manage resources. The major goal of healthcare logistics is to improve patient care and to maximize the use of the available resources. Also, the term "physician" has occasionally been replaced by "logistician." Today, operational research is regularly used to study policies, procedures, and trends related to the operation of the medical sector. Most frequently, operating managers of health and medical systems invoke the help of science in examining questions of priority rather than organization and decision support (Jimmy Triputra et al., 2023). It is important to stress that business can and should help hospitals in order to fulfill their mission. Logisticians and industrial engineers should be involved with the supply chain issues concerning the global healthcare sector. Numerous business possibilities exist by making the distribution of drugs, medical equipment, sanitation materials, and food to public and private hospitals more efficient and less expensive. Overall, advances in healthcare logistics improve patient care and reduce costs. Scientists in the sector argue that a person who goes to a hospital is a patient and a customer simultaneously (Hussain et al., 2023). A customer that should be satisfied, especially by not waiting for too long.

12.1 Collaboration with Medical Professionals

Introduction Any introduction to Operational Research (OR) work in healthcare has to reflect the peculiarities of that industry when compared to delivery and manufacturing. Those peculiarities generally take the form of the characteristics of any output from the production process, which in this case is the health, well-being, or treatment efficacy of the patient. In addition, the input to any production process is intangible and variable, changing even during the course of production. Patients can produce some of the outputs themselves and are crucial elements in the process (Balasubramanian et al., 2023). The resources available can be scarce, unevenly distributed, and could be increased or decreased by the patient's own behavior or decisions made by a third party. There may be ethical constraints or rules that are not openly available but are acted out by medical personnel. Clinical OR work has to be done with the collaboration of the people within the medical establishments, who, above anything else, have to maintain a level of trust about their professionalism and responsibility. This trust has to be held in the consulting OR professional, whether working in management structure or in one of the support services of the establishment, to enable his or her own topics of research to be taken fruitfully. In this, every member of the establishment provides a possible consultation and does consult an attached OR person (McDermott et al., 2021). Normally, empowerment is part of the gained benefit as a direct output from the systematic process of analysis.

12.2 Incorporating Public Health Principles

Public Health Principles All logistic coordination must ensure that public health principles for the prevention and treatment of illness and disability are never lost. For many serious life-threatening diseases, treatment and prevention programs are not optional. To forget this leads to the death of fragile individuals for lack of prevention and treatment that are all too often low-cost and available. When an infectious disease threatens calamity, the disease must be primary and income secondary for all. Active prevention must be strong if we are not to be buried under an exponential epidemic (Hallam & Contreras, 2018). This may mean a total mobilization of society going far beyond usual health care delivery. Everyone who suspects being infected must report and stay at home so as not to infect others. Activity must cease over an area if otherwise infecting migratory populations. It is obvious that cooperation between different countries and different non-governmental organizations is mandatory. Standard top-down organization is not acceptable; in many circumstances, it may be necessary to organize from the grassroots (Mwanza et al., 2023). In poor countries, sustaining emerging companies and jobs for all is less important and lower priority than sharing resources for everybody to have enough to eat, clean water, an education, and at least minimal health care. Companies founded on exploiting drug or supply monopolies for the gain of a few and the damage of many need to have limits set on their freedom so that the maximum number may benefit. Therefore, in a just society, access to essential drugs should not be in the gift of those who can pay but should be generally available at low cost, with marginal cost pricing only for non-essential items (Jha et al., 2022).

13 ETHICAL CONSIDERATIONS IN HEALTHCARE LOGISTICS

In recent years, healthcare has also been the focus of a growing amount of operations research applications due to key problems like gene sequencing, bed scheduling, and ambulance routing problems being part of it. However, despite these important achievements, the use of operations research in the healthcare field has to consider ethical considerations like no other. Probably, this is why only in the last couple of decades has optimization and simulation started to be applied to highly challenged problems like hospital management (Yu et al., 2021). Indeed, taking into account all the hard work that medical workers do, and considering that life is priceless, a single problem that ends up with the patient dying is enough to pretty much stop researchers from trying to pursue the path of solving these kinds of problems. The inclusion of these ethical considerations is very important since healthcare supply chain management has a direct impact on the quality of life of patients, where in most cases logistics are responsible for reducing patient recovery time. Indeed, operational aspects are greatly responsible for hospitals being expensive places to be. Calls to reduce healthcare costs can generate significant savings for taxpayers, meaning therefore that healthcare is a valuable and recognized research area, having also the advantage that healthcare provider organizations are considered ideal places to lead scientific research due to their data richness (Keskinocak & Savva, 2020; Rais & Vianaa, 2011).

13.1 Patient Privacy and Confidentiality

How do we share medical knowledge in a form that can benefit the community without putting at risk individual members' privacy and rights to confidentiality? Privacy is a human right, enshrined in national and international law. The risk of using data to the detriment of individuals has been exacerbated by electronic databases, modems, and other means of moving data in and out of the hospital or general practice (Minh Dang et al., 2019). Consequently, the rights of the patient to hold, protect, and control access to information about themselves have to be reconciled with the desirability of using patient databases for purposes beyond patient care. One way of making patient data less sensitive is by anonymizing the records. Data anonymization involves the removal of all personal or sensitive information from the records. Alas, there is no generally accepted technique for data anonymization and no commonly agreed legal test to say whether a set of records, modified as proposed by one technique, should be treated as anonymous. Consequently, a data custodian finds it difficult to define and respond to the difference between providing someone with a set of nominative records and providing an anonymous set of records (Nair et al., 2023). For the custodian, the difference may be in response times rather than technique or risk.

13.2 Equity and Access

Access has been talked about a lot in this section, but three further issues stand out. The first concerns inequities in performance - equity as a performance issue. Major variations in rates of adoption of health technologies and products remain across countries despite a simple existence proof that some countries are getting their health service constituent parts to function better than others. Of course, some variation should be expected, as differences in aggregate health spending, incidence of different diseases, population structure, and cultural values can all influence how many procedures are undertaken. Nonetheless, large variations in the rate of implementation of specific effective products across similar populations, when there appears to be no 'good reason' for such differences, point logically to local management/regulatory as the main contributing factor (Jha et al., 2022). Are current providers of service really providing value for the corresponding recipients of service?

14 REGULATORY COMPLIANCE IN HEALTHCARE LOGISTICS

Just like with product flows, logistics processes for healthcare flows should consider regulatory aspects. In healthcare logistics, ensuring and protecting patients must be the most critical issues, and such high objectives make healthcare logistics more complicated compared to pure business logistics. Government regulations on healthcare are strict to enable secure service to patients and prevent them from informational and physical harm. Security and safety investments in healthcare have favorable but different and complex relationships with logistics-related benefits (Mahjoob et al., 2023). Therefore, while taking advantage of the developed methods in business logistics, healthcare logistics has more varied operations due to healthcare specifics. In healthcare logistics, customizing logistics policies for regulatory compliance is complicated and vital. While managing and increasing compliance with rules, guidelines, standards, laws, and protocols, it is essential to minimize healthcare logistics operational costs such as inventory and expiry costs. The key is to proceed in parallel, both directly and indirectly, while enhancing the quality of service. Most of the methods in the chapter on regulations standardize the effectiveness of events and express the detailed application of the presented approaches in healthcare logistics (Dion & Evans, 2024). They work for both the physical properties of goods, such as temperature, and the delivery of information goods, such as personal information.

14.1 Legal Frameworks and Requirements

Public Health Service was created in 1946 within the FATA by incorporating a variety of institutions and services such as the Queen's Clinic, vaccination, disciplinary court, the Central Department of Health, Old Military Pavilion, and the Mobile Clinic, among others. The Panamanian Social Security Fund and the Board of Trustees of the Health Information and Statistics System were created in 1941. The Dr. Arnulfo Arias Madrid Hospital Complex incorporates specialized health care services at various levels of complexity, as well as the social security fund hospital in maternity, pediatrics, psychiatry, oncology, plastic surgery, general surgery, thoracic surgery, among others, as well as three hospital annexes; the Psychiatric Hospital of Panama; the Institute of Physical Medicine as well as a network of polyclinics within the Metropolitan Area, the interior, and outside the country (Keskinocak & Savva, 2020). On one side, there are a set of legal obligations and requirements that budgets always invest in health infrastructure and state assets, but it takes a long time to implement and is sometimes not within a reasonable timeframe. The state, through the different administrative bodies with jurisdiction requires other health facilities and services, which it does not have and is forced to acquire through the private sector by signing contracts or proposals to request pensions from interested investors. This legal framework requires the regulatory authorities to adequately control the economic provisions that have been implemented in public service, in financial resources as well as in the application of contracts, which is not always possible (Farahani et al., 2023).

14.2 Compliance Strategies

We consider enforcing compliance through rewards and monitoring. The compliance of a service system to a quality of service may be improved through rewards, such as bonuses or penalty reductions in the case of public service contracts, and monitoring. With T being the time needed to inspect a server for non-compliance, r being the effectiveness of the server, q being the penalty function in the case of non-compliance, and C being the budget limit, the problem is to allocate rewards to servers over time intervals and to select servers for inspection during each interval, so as to minimize the expected penalty cost, minus the rewards spent, over the corresponding inspection subintervals plus the costs of inspection (Detwal et al., 2024). This class of models is a generalization of personnel planning models in which penalties have not been specifically introduced in the past as a criterion to be optimized. In the staff allocation model, personnel are to be allocated to shifts over a finite planning horizon with the objective of minimizing the overall penalty costs. The penalty costs are incident upon events of understaffing or overstaffing and of preferences being violated (Saville et al., 2019). These events would lead to under- or over-utilization of available resources, forbidden operating conditions, complications with the execution of tasks, and a possible decrease in the quality of offered services. The basic model considers the specific case in which the staff of each type are designed to cover exactly one shift each day and in which the available personnel of each type are defined over each shift of each day. The more general model assumes rostering opportunities with the personnel of each type allocated to some sets of consecutive shifts of the planning horizon. The cost of staff allocation is the sum of the penalty cost and the cost that the labor contracts mandate (Joshi et al., 2022).

15 FUTURE TRENDS AND INNOVATIONS IN HEALTHCARE LOGISTICS

There are some innovative features of health care organizations, which are emerging and will shape the future of healthcare logistics. Supply chain coordination: Greater coordination among different participants in the logistics system, including healthcare organizations, would provide a better supply chain in health care, which will increase healthcare quality. Food and pharmacy plaza: An increasing number of hospitals are adopting the concept of pharmacies and nutrition or food departments and connecting them. By doing this, they are able to meet special delivery requirements and decrease delivery costs (Saraswat et al., 2022). Service-oriented logistics: Healthcare organizations increasingly use outsourced services for their support and caregiving activities. RFID and RTLS technology: An increasing number of hospitals are adopting RFID and RTLS technology, which will keep track of the equipment and employees and reduce management efforts and lost equipment costs. Using robots for automatic transportation and delivery: Robots are being implemented to deliver medication and laboratory specimens in hospitals. Some hospitals use these transportation robots to supply patient care items, such as pillows and blankets, to nursing units 24/7 (Ramalingam et al., 2024). The underground mail system, if it is viable to be implemented, is able to achieve faster delivery times. Multilevel loading areas, like the ones used in warehouses, could be implemented on various floors, thus speeding loading times. Postponement of important logistics activities and infrastructures: Nowadays, many hospitals are considering or even implementing an automatic restocking system, where the stocks are monitored and refilled with the required items by the manufacturers themselves (Lamé et al., 2022). There is a possibility that large areas are reserved solely for unloading/loading operations of trucks.

Service megacenter: If large logistic service centers were established regionally, their services would make it easier for hospitals to connect with external logistic units. By combining transportation for multiple hospitals, economies of scale could be achieved. A wastewater heat recovery system: The building incorporates a wastewater heat recovery system that captures waste heat from the outpatient clean space and uses it to heat the domestic hot water. The actual implementation of some of the above-mentioned innovations could prove to be difficult since each healthcare facility is unique (Kraus et al., 2021). Factors like space availability, capital investment, and interconnection among different subsystems will play a critical role in the performance of the proposed logistic system. The growing size of the healthcare system also acts as a restraining factor for the actual implementation of these systems. Although these engineering solutions could seem viable at one hospital, because of varying constraints, like geography and size of the hospital, the allowance for these initiatives differs accordingly, thus making generalization of the actual intervention difficult to articulate (Rais & Vianaa, 2011). Despite these challenges, over the next few decades, complex healthcare systems stand to benefit considerably from the adoption of integrated engineering systems to achieve efficient healthcare logistics.

15.1 Predictive Analytics and Forecasting

Predictive analytics evaluates the data or content of the past to make assessments about the future. Although the assessment is built on data, the data conveys a story about the direction the organization may want to move. Both service providers and health care professionals are very interested in reducing no-shows and cancellations of medical appointments. Several approaches have been suggested to predict patient arrivals in clinics; particularly, it casts an uncertain factor in determining the proportion of patients who exactly attend their appointments (Kakalou et al., 2018). These approaches include analyzing past patient arrival data, identifying factors influencing patient arrival rates, using scheduling policies to decrease patient running and waiting times, and using simulation techniques to forecast patient flow distributions. Several managerial recommendations are formulated, highlighting that the major factor in determining patient attendance is the postponement of appointments. Furthermore, management must clarify the particular strengths and weaknesses of the developed model, which can be built upon more tests and adapted to changing medical disciplines (Dion & Evans, 2024). The strength of using technology and data analytics allows the forecast of patient non-attendance with greater accuracy, reduces organizational costs, matches demand and supply, reduces patient waiting times, and improves health care systems. Indeed, forecasting patient non-attendance by mining big datasets could heavily affect the resilience of the system, which in turn provides a series of relevant tactical or strategic actions aimed at coordinating adaptive choices. In essence, by knowing more about patients' possible reasons for missing appointments, building in protections to help patients overcome those barriers, and finding ways to have greater flexibility in provider schedules can result in improved access to healthcare (Rais & Vianaa, 2011).

15.2 Automation and Robotics

The use of both automation and robotics can be a means through which the demands of improving efficiency in service delivery while maintaining high levels of or even improving the quality of healthcare services can be met. Surgery, pharmacy, and general services could gain a number of advantages from the application of this type of technology, while the patient could gain from the likely improvements in clinical procedures. The complexity is what limits a greater take-up, although this is likely to reduce at some stage as technology advances and the costs of the equipment fall (Detwal et al., 2024). The main text discusses these issues in detail, looking first at the areas and types in surgical applications, overhead and room delivery, and then at both hospital pharmacy and its staffing issues. The following sections focus on the use of robotics in pharmacy automation where the main focuses are on systems with sophisticated robotic technology that can perform a full range of medication inventory management and dispensing functions. Even though robots are unlikely to completely replace human pharmacists, the processes at dispense counters and drug distribution points will be quickly superseded as systems continue to evolve and reduce their costs (Leite et al., 2021). The growing number of pharmacies benefiting from the new technology suggests robots are poised to revolutionize the industry.

15.3 Summary of Key Findings

Throughout this review, it is evident that the development of operations research, especially during the past 20 years, has enabled an increasingly sophisticated use of these methods within healthcare logistics and transportation. The size, complexity, and apparent intractability of many of the problems addressed suggest that there are numerous opportunities to add further value in both methodological development and practical applications (Joshi et al., 2022). These include the use of real-time, embedded, and sensor-driven data gathering combined with classical operations research approaches, increasingly intelligent and autonomous systems that are more suited to the wide variety of real operating conditions that are seen, and better management of externalities such as air and noise pollution. The goal of most of the work presented here has been to better understand the use of complex healthcare resources with the aim of improving healthcare supply chains to improve outcomes for patients while also improving the efficiency and effectiveness of the healthcare providers (Ilangakoon et al., 2022). Indeed, healthcare can provide a rich and fertile application area where operational researchers can apply their skills to socially important and technically demanding problems that require the management of a range of complex, often partly conflicting, performance metrics. We conclude that this review has barely scratched the surface of the issues and opportunities offered. As the range of specialist skills that contribute to capacity in such operations research problems continues to widen, it is likely that real-world impact will demonstrably increase (Diwas Singh et al., 2020).

15.4 Research Gaps and Areas for Future Exploration

Given the considerable benefits and potential effectiveness of operational research in the healthcare sector, there is a surprising paucity of studies in the current literature. This also offers significant potential for researchers in the field of transport and logistics to contribute to the efficiency, and more importantly, the effectiveness of the healthcare sector. We call to stage a significant increase of research in this sector. There is a requirement for not only the development and implementation of more effective tactical and operational management tools, but also strategic management and conceptual tools which impact the conflicts of interest that result from different economic activities in the sector (Kraus et al., 2021). These conflicts of interest may often not be economic in nature but more ethical. This chapter has shown a number of possible encounters by investigating over twenty papers that address both operational research and the healthcare sector, and there is potential for researchers hailing from the fields of transport and logistics to make significant contributions to this relatively virgin sector. The strategic long-term view presented in this chapter leads to the following conclusions of future research. It is clear that one of the key rationales of the healthcare sector is to cure and to reduce suffering of humans (Hussain et al., 2023). Therefore, we expect that, as current practice has led, operational research models for decision support, predominantly but not exclusively from the fields of transportation, logistics, and supply chain management, will advance. All this will be accomplished with the aim to contribute to a more effective and efficient healthcare sector. Given a solid experience from important application areas in these fields, we further believe that future challenges, in particular from the qualitative, customer-oriented perspective (including patients, service providers, payers, and governments), will also be addressed successfully. In balancing trade-offs in a wide range of quantitative as well as qualitative, strategic, tactical, and operational healthcare management situations, operational research provides invaluable decision support (Marsilio et al., 2022). As the contributions in this chapter illustrate, these efforts are continuing and future studies are invited to investigate these challenging and vastly rewarding areas of operational research.

16 CONCLUSION

In modern societies, the healthcare system is an area of strategic importance. A significant part of healthcare costs is associated with the transportation of patients and the ingoing and outgoing flows of hospital logistics. Micrologistic problems derived from these flows are crucial and are involved in the quest for economic and social improvement of the healthcare system. In view of the increasing importance of the problems related to logistics and transport in the healthcare system, this paper presents a review of the current literature on operational research, including transport and logistics applied to the healthcare sector. Here, different facets of this problem are presented, giving special emphasis to the practical relevance of the addressed problems and to the adopted methodologies. This last section includes the conclusions and future directions. The section is a brief conclusion. The last section includes the final conclusions and presents future directions for research, based on the main limitations of the review and the opportunities identified. The opportunities were identified so that the operational research community can better help healthcare structures to improve their organization and expenditure. Indeed, the healthcare sector poses a wide variety of disrupting and challenging problems to be addressed, ranging from the ledgering of resources to highly sophisticated surgery, where it is crucial to minimize reaction while optimizing success. Finally, the suggestions provided in this section are aimed at both researchers and policy-makers who are interested in the related fields.

REFERENCES

Alolayyan, M. N., & Alyahya, M. S. (2023). Operational flexibility impact on hospital performance through the roles of employee engagement and management capability. *BMC Health Services Research, 23*(1), 19. Advance online publication. DOI: 10.1186/s12913-023-09029-y PMID: 36624513

Balasubramanian, S., Shukla, V., Islam, N., Upadhyay, A., & Duong, L. (2023). Applying artificial intelligence in healthcare: Lessons from the COVID-19 pandemic. *International Journal of Production Research*. Advance online publication. DOI: 10.1080/00207543.2023.2263102

Bhat, S., Gijo, E. V., Antony, J., & Cross, J. (2023). Strategies for successful deployment and sustainment of Lean Six Sigma in healthcare sector in India: A multi-level perspective. *The TQM Journal, 35*(2), 414–445. Advance online publication. DOI: 10.1108/TQM 10 2021 0302

Cannavale, C., Esempio Tammaro, A., Leone, D., & Schiavone, F. (2022). Innovation adoption in inter-organizational healthcare networks – the role of artificial intelligence. In *European Journal of Innovation Management* (Vol. 25, Issue 6). DOI: 10.1108/EJIM-08-2021-0378

Carter, M. W., & Busby, C. R. (2023). How can operational research make a real difference in healthcare? Challenges of implementation. *European Journal of Operational Research, 306*(3), 1059–1068. Advance online publication. DOI: 10.1016/j.ejor.2022.04.022

Crowe, S., & Utley, M. (2022). Praxis in healthcare OR: An empirical behavioural OR study. *The Journal of the Operational Research Society, 73*(7), 1444–1456. Advance online publication. DOI: 10.1080/01605682.2021.1919036

Detwal, P. K., Agrawal, R., Samadhiya, A., Kumar, A., & Garza-Reyes, J. A. (2024). Revolutionizing healthcare organizations with Operational Excellence and Healthcare 4.0: a systematic review of the state-of-the-art literature. In *International Journal of Lean Six Sigma* (Vol. 15, Issue 1). DOI: 10.1108/IJLSS-04-2023-0061

Dion, H., & Evans, M. (2024). Strategic frameworks for sustainability and corporate governance in healthcare facilities; approaches to energy-efficient hospital management. *Benchmarking, 31*(2), 353–390. Advance online publication. DOI: 10.1108/BIJ-04-2022-0219

Diwas Singh, K. C., Scholtes, S., & Terwiesch, C. (2020). Empirical research in healthcare operations: Past research, present understanding, and future opportunities. *Manufacturing & Service Operations Management, 22*(1), 73–83. Advance online publication. DOI: 10.1287/msom.2019.0826

Dodge, J. R., Youles, B., Caldararo, J., Sears, E. D., Caverly, T. J., Michael Ho, P., Shimada, S. L., Kaboli, P., Albright, K., Robinson, S. A., McNeal, D. M., Damschroder, L., Saini, S. D., & Adams, M. A. (2023). Engaging Operational Partners Is Critical for Successful Implementation of Research Products: A Coincidence Analysis of Access-Related Projects in the Veterans Affairs Healthcare System. *Journal of General Internal Medicine, 38*(S3), 923–930. Advance online publication. DOI: 10.1007/s11606-023-08115-5 PMID: 37340262

Dzando, G., Salifu, S., Donyi, A. B., Akpeke, H., Kumah, A., Dordunu, R., & Nonoh, E. A. (2022). Healthcare in Ghana amidst the coronavirus pandemic: A narrative literature review. In *Journal of Public Health Research* (Vol. 11, Issue 1). DOI: 10.4081/jphr.2021.2448

Farahani, R. Z., Ruiz, R., & Van Wassenhove, L. N. (2023). Introduction to the special issue on the role of operational research in future epidemics/ pandemics. In *European Journal of Operational Research* (Vol. 304, Issue 1). DOI: 10.1016/j.ejor.2022.07.019

Fitzpatrick, G., Decroo, T., Draguez, B., Crestani, R., Ronsse, A., Van den Bergh, R., & Van Herp, M. (2017). Operational research during the Ebola emergency. *Emerging Infectious Diseases, 23*(7), 1057–1062. Advance online publication. DOI: 10.3201/eid2307.161389 PMID: 28628463

Fusco, F., Marsilio, M., & Guglielmetti, C. (2023). Co-creation in healthcare: framing the outcomes and their determinants. In *Journal of Service Management* (Vol. 34, Issue 6). DOI: 10.1108/JOSM-06-2021-0212

Hallam, C. R. A., & Contreras, C. (2018). Lean healthcare: scale, scope and sustainability. In *International Journal of Health Care Quality Assurance* (Vol. 31, Issue 7). DOI: 10.1108/IJHCQA-02-2017-0023

Hussain, M., Ajmal, M., Subramanian, G., Khan, M., & Anas, S. (2023). Challenges of big data analytics for sustainable supply chains in healthcare – a resource-based view. *Benchmarking*. Advance online publication. DOI: 10.1108/BIJ-06-2022-0390

Ilangakoon, T. S., Weerabahu, S. K., Samaranayake, P., & Wickramarachchi, R. (2022). Adoption of Industry 4.0 and lean concepts in hospitals for healthcare operational performance improvement. *International Journal of Productivity and Performance Management, 71*(6), 2188–2213. Advance online publication. DOI: 10.1108/IJPPM-12-2020-0654

Jha, A., Athanerey, A., & Kumar, A. (2022). Role and challenges of internet of things and informatics in Healthcare research. In *Health and Technology* (Vol. 12, Issue 4). DOI: 10.1007/s12553-022-00661-y

Joshi, S., Sharma, M., Das, R. P., Rosak-Szyrocka, J., Żywiołek, J., Muduli, K., & Prasad, M. (2022). Modeling Conceptual Framework for Implementing Barriers of AI in Public Healthcare for Improving Operational Excellence: Experiences from Developing Countries. *Sustainability (Basel), 14*(18), 11698. Advance online publication. DOI: 10.3390/su141811698

Kakalou, E., Riza, E., Chalikias, M., Voudouri, N., Vetsika, A., Tsiamis, C., Choursoglou, S., Terzidis, A., Karamagioli, E., Antypas, T., & Pikoulis, E. (2018). Demographic and clinical characteristics of refugees seeking primary healthcare services in Greece in the period 2015-2016: A descriptive study. *International Health, 10*(6), 421–429. Advance online publication. DOI: 10.1093/inthealth/ihy042 PMID: 29992276

Keskinocak, P., & Savva, N. (2020). A review of the healthcare-management (modeling) literature published in manufacturing & service operations management. *Manufacturing & Service Operations Management, 22*(1), 59–72. Advance online publication. DOI: 10.1287/msom.2019.0817

Kraus, S., Schiavone, F., Pluzhnikova, A., & Invernizzi, A. C. (2021). Digital transformation in healthcare: Analyzing the current state-of-research. *Journal of Business Research, 123*, 557–567. Advance online publication. DOI: 10.1016/j.jbusres.2020.10.030

Lamé, G., Crowe, S., & Barclay, M. (2022). "What's the evidence?"—Towards more empirical evaluations of the impact of OR interventions in healthcare. *Health Systems (Basingstoke, England), 11*(1), 59–67. Advance online publication. DOI: 10.1080/20476965.2020.1857663 PMID: 35127059

Lamé, G., Crowe, S., Komashie, A., & Royston, G. (2023). Joining forces: The value of design partnering with operational research to improve healthcare delivery. *Design Science, 9*, e4. Advance online publication. DOI: 10.1017/dsj.2023.2

Leite, H., Lindsay, C., & Kumar, M. (2021). COVID-19 outbreak: Implications on healthcare operations. *The TQM Journal*, *33*(1), 247–256. Advance online publication. DOI: 10.1108/TQM-05-2020-0111

Leone, D., Schiavone, F., Appio, F. P., & Chiao, B. (2021). How does artificial intelligence enable and enhance value co-creation in industrial markets? An exploratory case study in the healthcare ecosystem. *Journal of Business Research*, *129*, 849–859. Advance online publication. DOI: 10.1016/j.jbusres.2020.11.008

Mahjoob, A., Alfadhli, Y., & Omachonu, V. (2023). Healthcare Waste and Sustainability: Implications for a Circular Economy. *Sustainability (Basel)*, *15*(10), 7788. Advance online publication. DOI: 10.3390/su15107788

Marsilio, M., Pisarra, M., Rubio, K., & Shortell, S. (2022). Lean adoption, implementation, and outcomes in public hospitals: Benchmarking the US and Italy health systems. *BMC Health Services Research*, *22*(1), 122. Advance online publication. DOI: 10.1186/s12913-022-07473-w PMID: 35090455

McDermott, O., Antony, J., & Douglas, J. (2021). Exploring the use of operational excellence methodologies in the era of COVID-19: Perspectives from leading academics and practitioners. *The TQM Journal*, *33*(8), 1647–1665. Advance online publication. DOI: 10.1108/TQM-01-2021-0016

Minh Dang, L., Piran, M. J., Han, D., Min, K., & Moon, H. (2019). A survey on internet of things and cloud computing for healthcare. *Electronics (Basel)*, *8*(7), 768. Advance online publication. DOI: 10.3390/electronics8070768

Mwanza, J., Telukdarie, A., & Igusa, T. (2023). Impact of industry 4.0 on healthcare systems of low- and middle- income countries: a systematic review. In *Health and Technology* (Vol. 13, Issue 1). DOI: 10.1007/s12553-022-00714-2

Nair, D., Thekkur, P., Fernando, M., Kumar, A. M. V., Satyanarayana, S., Chandraratne, N., Chandrasiri, A., Attygalle, D. E., Higashi, H., Bandara, J., Berger, S. D., & Harries, A. D. (2023). Outcomes and Challenges in Noncommunicable Disease Care Provision in Health Facilities Supported by Primary Health Care System Strengthening Project in Sri Lanka: A Mixed-Methods Study. *Healthcare (Basel)*, *11*(2), 202. Advance online publication. DOI: 10.3390/healthcare11020202 PMID: 36673570

Palmer, R., Fulop, N. J., & Utley, M. (2018). A systematic literature review of operational research methods for modelling patient flow and outcomes within community healthcare and other settings. *Health Systems (Basingstoke, England)*, *7*(1), 29–50. Advance online publication. DOI: 10.1057/s41306-017-0024-9 PMID: 31214337

Patel, J., More, S., Sohani, P., Bedarkar, S., Dinesh, K. K., Sharma, D., Dhir, S., Sushil, S., & Ghosh, R. S. (2024). Reshaping the equitable and inclusive access to healthcare: A qualitative study. *Clinical Epidemiology and Global Health*, *26*, 101544. Advance online publication. DOI: 10.1016/j.cegh.2024.101544 PMID: 38707587

Pearson, M., Monks, T., Gibson, A., Allen, M., Komashie, A., Fordyce, A., Harris-Golesworthy, F., Pitt, M. A., Brailsford, S., & Stein, K. (2013). Involving patients and the public in healthcare operational research-The challenges and opportunities. *Operations Research for Health Care*, *2*(4), 86–89. Advance online publication. DOI: 10.1016/j.orhc.2013.09.001

Pradhan, B., Bhattacharyya, S., & Pal, K. (2021). IoT-Based Applications in Healthcare Devices. In *Journal of Healthcare Engineering* (Vol. 2021). DOI: 10.1155/2021/6632599

Rais, A., & Vianaa, A. (2011). Operations research in healthcare: A survey. *International Transactions in Operational Research*, *18*(1), 1–31. Advance online publication. DOI: 10.1111/j.1475-3995.2010.00767.x

Ramalingam, S., Subramanian, M., Sreevallabha Reddy, A., Tarakaramu, N., Ijaz Khan, M., Abdullaev, S., & Dhahbi, S. (2024). Exploring business intelligence applications in the healthcare industry: A comprehensive analysis. *Egyptian Informatics Journal*, *25*, 100438. Advance online publication. DOI: 10.1016/j.eij.2024.100438

Sachdeva, R., Williams, T., & Quigley, J. (2007). Mixing methodologies to enhance the implementation of healthcare operational research. *The Journal of the Operational Research Society*, *58*(2), 159–167. Advance online publication. DOI: 10.1057/palgrave.jors.2602293

Saraswat, D., Bhattacharya, P., Verma, A., Prasad, V. K., Tanwar, S., Sharma, G., Bokoro, P. N., & Sharma, R. (2022). Explainable AI for Healthcare 5.0: Opportunities and Challenges. In *IEEE Access* (Vol. 10). DOI: 10.1109/ACCESS.2022.3197671

Saville, C. E., Smith, H. K., & Bijak, K. (2019). Operational research techniques applied throughout cancer care services: a review. In *Health Systems* (Vol. 8, Issue 1). DOI: 10.1080/20476965.2017.1414741

Shafik, W. (2024a). Artificial Intelligence-Enabled Internet of Medical Things (AIoMT) in Modern Healthcare Practices. In *Clinical Practice and Unmet Challenges in AI-Enhanced Healthcare Systems* (pp. 42–69). IGI Global. DOI: 10.4018/979-8-3693-2703-6.ch003

Shafik, W. (2024b). Artificial Intelligence-Enabled Internet of Medical Things for Enhanced Healthcare Systems. In *Smart Healthcare Systems* (pp. 119–134). CRC Press., DOI: 10.1201/9781032698519-9

Shafik, W. (2024c). Connected healthcare—the impact of Internet of Things on medical services. In *Artificial Intelligence and Internet of Things based Augmented Trends for Data Driven Systems* (pp. 181–217). CRC Press., DOI: 10.1201/9781003497318-10

Shafik, W. (2024d). Data-Driven Future Trends and Innovation in Telemedicine. In *Improving Security, Privacy, and Connectivity Among Telemedicine Platforms* (pp. 93–118). IGI Global. DOI: 10.4018/979-8-3693-2141-6.ch005

Shafik, W. (2024e). Digital healthcare systems in a federated learning perspective. In *Federated Learning for Digital Healthcare Systems* (pp. 1–35). Elsevier. DOI: 10.1016/B978-0-443-13897-3.00001-1

Shafik, W. (2024f). Digital healthcare systems in a federated learning perspective. In *Federated Learning for Digital Healthcare Systems* (pp. 1–35). Elsevier., DOI: 10.1016/B978-0-443-13897-3.00001-1

Shafik, W. (2024g). Healthcare and Its Applications in a Smart City. In *Healthcare-Driven Intelligent Computing Paradigms to Secure Futuristic Smart Cities* (pp. 1–35). Chapman and Hall/CRC. DOI: 10.1201/9781032631738-1

Shafik, W. (2024h). IoMT Future Trends and Challenges: Emerging Technologies, Policy Implications, and Research Questions. *Lightweight Digital Trust Architectures in the Internet of Medical Things (IoMT)*, 348–370.

Shafik, W. (2024i). IoT-Enabled Secure and Intelligent Smart Healthcare. In *Secure and Intelligent IoT-Enabled Smart Cities* (pp. 308–333). IGI Global., DOI: 10.4018/979-8-3693-2373-1.ch015

Shafik, W. (2024j). IoT-Enabled Secure and Intelligent Smart Healthcare: Beyond 5G in Enabling Smart Cities. In *Secure and Intelligent IoT-Enabled Smart Cities* (pp. 308–333). IGI Global.

Shafik, W. (2024k). Smart Health Revolution: Exploring Artificial Intelligence of Internet of Medical Things. In *Healthcare Industry Assessment: Analyzing Risks, Security, and Reliability* (pp. 201–229). Springer.

Shafik, W. (2024l). Smart Health Revolution: Exploring Artificial Intelligence of Internet of Medical Things. In *Healthcare Industry Assessment: Analyzing Risks, Security, and Reliability* (pp. 201–229). Springer. DOI: 10.1007/978-3-031-65434-3_9

Shafik, W. (2024m). The Future of Healthcare: AIoMT—Redefining Healthcare with Advanced Artificial Intelligence and Machine Learning Techniques. In *Artificial Intelligence and Machine Learning in Drug Design and Development* (pp. 605–634). Wiley., DOI: 10.1002/9781394234196.ch19

Shafik, W. (2024n). The Future of Healthcare: Data-Driven Trends and Innovation. In *Advancing Healthcare through Data-driven Innovations* (pp. 166–185). CRC Press. DOI: 10.1201/9781003475392-11

Shafik, W. (2024o). Toward a More Ethical Future of Artificial Intelligence and Data Science. In *The Ethical Frontier of AI and Data Analysis* (pp. 362–388). IGI Global. DOI: 10.4018/979-8-3693-2964-1.ch022

Shafik, W. (2025a). Artificial Intelligence Assisted Internet of Medical Things (AIoMTs) in Sustainable Healthcare Ecosystem. *Wellness Management Powered by AI Technologies*, 75–101.

Shafik, W. (2025b). Revolutionizing Skin Cancer Diagnosis With Artificial Intelligence: Insights Into Machine Learning Techniques. In *Impact of Digital Solutions for Improved Healthcare Delivery* (pp. 167–194). IGI Global.

Shafik, W. (2025c). Revolutionizing Skin Cancer Diagnosis With Artificial Intelligence: Insights Into Machine Learning Techniques. In *Impact of Digital Solutions for Improved Healthcare Delivery* (pp. 167–194). IGI Global.

Shafik, W. (2025d). Smart biomedical devices for smart healthcare. In *Machine Learning Models and Architectures for Biomedical Signal Processing* (pp. 421–448). Elsevier. DOI: 10.1016/B978-0-443-22158-3.00017-X

Shafik, W. (2025e). Smart biomedical devices for smart healthcare. In *Machine Learning Models and Architectures for Biomedical Signal Processing* (pp. 421–448). Elsevier. DOI: 10.1016/B978-0-443-22158-3.00017-X

Shafk, W. (2023). Wearable Medical Electronics in Artifcial Intelligence of Medical Things. In *Handbook of Security and Privacy of AI-Enabled Healthcare Systems and Internet of Medical Things*. https://doi.org/DOI: 10.1201/9781003370321-2

Singh, A., & Parida, R. (2022). Decision-Making Models for Healthcare Supply Chain Disruptions: Review and Insights for Post-pandemic Era. *International Journal of Global Business and Competitiveness*, *17*(2), 130–141. Advance online publication. DOI: 10.1007/s42943-021-00045-5

Sordan, J. E., Marinho, C. A., Oprime, P. C., Pimenta, M. L., & Andersson, R. (2023). Characterization of Lean Six Sigma projects in healthcare settings: Empirical research. *Benchmarking*, *30*(10), 4058–4075. Advance online publication. DOI: 10.1108/BIJ-03-2022-0183

Syahrir, I., Suparno, , & Vanany, I. (2015). Healthcare and Disaster Supply Chain: Literature Review and Future Research. *Procedia Manufacturing*, *4*, 2–9. Advance online publication. DOI: 10.1016/j.promfg.2015.11.007

Tippong, D., Petrovic, S., & Akbari, V. (2022). A review of applications of operational research in healthcare coordination in disaster management. In *European Journal of Operational Research* (Vol. 301, Issue 1). DOI: 10.1016/j.ejor.2021.10.048

Triputra, J., Nurhayati, N., Choirunisa, R., Pamungkasari, T., & Paramata, V. (2023). Peningkatan Efektivitas Manajemen Kesehatan Melalui Implementasi Teknologi Informasi. *Jurnal Ilmiah Kedokteran Dan Kesehatan*, *3*(1), 220–225. Advance online publication. DOI: 10.55606/klinik.v3i1.2319

Ur Rasool, R., Ahmad, H. F., Rafique, W., Qayyum, A., Qadir, J., & Anwar, Z. (2023). Quantum Computing for Healthcare: A Review. In *Future Internet* (Vol. 15, Issue 3). DOI: 10.3390/fi15030094

Walsh, M., Kittler, M. G., & Mahal, D. (2018). Towards a new paradigm of healthcare: Addressing challenges to professional identities through Community Operational Research. *European Journal of Operational Research*, *268*(3), 1125–1133. Advance online publication. DOI: 10.1016/j.ejor.2017.05.052

Wiig, S., Aase, K., Billett, S., Canfield, C., Røise, O., Njå, O., Guise, V., Haraldseid-Driftland, C., Ree, E., Anderson, J. E., Macrae, C., Bourrier, M., Berg, S. H., Bergerød, I. J., Schibevaag, L., Øyri, S. F., Sjøseth, S., O'Hara, J., Kattouw, C. E., . . . Jeppesen, E. (2020). Defining the boundaries and operational concepts of resilience in the resilience in healthcare research program. In *BMC Health Services Research* (Vol. 20, Issue 1). DOI: 10.1186/s12913-020-05224-3

Yu, W., Zhao, G., Liu, Q., & Song, Y. (2021). Role of big data analytics capability in developing integrated hospital supply chains and operational flexibility: An organizational information processing theory perspective. *Technological Forecasting and Social Change*, *163*, 120417. Advance online publication. DOI: 10.1016/j.techfore.2020.120417

Chapter 7
Managing Educational Natural Language Processing Projects With Hybrid Artificial Intelligence:
A Case Study in Arabic Text Simplification

Naoual Nassiri
https://orcid.org/0000-0002-8299-9422
Engineering and Sustainable Development Team, Higher School of Technology, Ibnou Zohr University, Dakhla, Morocco

Safae Berrichi
University Mohamed I, Oujda, Morocco

Azzeddine Mazroui
https://orcid.org/0000-0001-8791-0055
University Mohamed I, Oujda, Morocco

Abdelhak Lakhouaja
University Mohamed I, Oujda, Morocco

ABSTRACT

We present a case study in the management of an educational research project through the design of an Arabic text simplification system. The push towards AI

DOI: 10.4018/979-8-3373-2975-8.ch007

implementations within education has produced a parallel demand for personalized, accessible content to accommodate language learners. This research project builds towards addressing these gaps, by designing and developing a lexical simplification framework that combines linguistic resources with the latest advances in deep learning models. It takes the reader through the project lifecycle while also noting the overall decisions, methodological innovations and management strategies made throughout the project lifecycle. The system produces identifications of complex words and simplified synonym options based on an innovative combination of the Arabic WordNet, lexicons manually composed, word embeddings and the pre-trained AraBERT model. The outcomes are assessed using a combination of text readability classification and semantic similarity, to produce quantifiable improvements in readability while sufficiently retaining meaning.

1. INTRODUCTION

The quick expansion of Artificial Intelligence (AI)-focused learning tools promptly transforms the approaches to language learning and access to study materials. Increasingly, AI technologies are being incorporated into existing platforms for learning, enabling flexible content, learners' needs assessment and integration. Among the new technologies is automatic text simplification, which has also shown itself to be an extremely promising tool for diminishing linguistic obstacles, thereby allowing complex texts to be made more readable to more learners. These include second-language learners (L2), reading-disabled students, and learners with cognitive disabilities.

Text Simplification facilitates not only learner's personalization of learning experience, but also equality of education, which facilitates inclusive learning environments that allow learners to engage with information at levels that are meaningful to their linguistic proficiency level. Although most research in this domain has gone towards well-endowed languages like English, the resource-scarce language with rich morphology like Arabic has not been exposed to similar amounts of attention in automatic text simplification tool research. Natural Language Processing (NLP) efforts in education within the domain of Arabic text simplification can potentially contribute a lot to expanded access for Arabic learners, allowing educational equity and inclusion in regions where Arabic is spoken.

It is this chapter's focus to deal with the management of educational NLP projects, in this instance, Arabic text simplification through lexical simplification with a view to replacing complicated words with easy ones, without changing the grammatical form and sense of the text. The project is one of urgent pedagogical needs, simplification of actual Arabic texts in a range of environments. To do this, the research project used a hybrid approach, one which mixed traditional linguistic materials with modern deep learning models. More specifically, it mixed synonym lexicons and morphological resources with word embeddings and pretrained Large Language Models (LLMs) (Rhel & Roussinov, 2025). Merging both pre-set syntactic methods and data-driven methods allowed the project to overcome the limitations of existing methodologies, while keeping the strengths of both intact.

Design, implementation, and evaluation of the project were conducted under a research management framework, displaying critical features of educational project planning, resource management, data management, model evaluation, and stakeholder engagement. Through examination of the technical and managerial facets of the research, the chapter offers a model for integrating successful AI-initiated projects within education innovation agendas. Table 1 presents the project roadmap.

Table 1. Project roadmap

Phase	Description
Planning & Team Setup	Defined pedagogical objectives, identified learner needs, and assigned roles (linguists, NLP engineers, coordinators).
Corpus Construction	Compiled and annotated 819 authentic Arabic learning texts, creating the foundation for simplification.
System Development	Designed a hybrid simplification pipeline combining linguistic resources with AI models, developed iteratively through short Agile sprints.
Evaluation	Assessed outputs using readability metrics, semantic similarity (BERTScore), and teacher validation to ensure pedagogical relevance.
Deployment & Future Use	Outlined hosting, maintenance, and integration into educational platforms for sustainable, real-world use.

2. CONTEXT AND AIMS OF THE PROJECT: BETWEEN PEDAGOGIC NEEDS AND LINGUISTIC CONSTRAINTS

a. Origins of the project

The project is one of a lively digital move towards teaching, centered on equity and access to learning. More particularly, it is a response to a problem many experiences in the learning of Arabic as a foreign language: the disparity between the degree of linguistic complexity of available texts and the learners' effective level.

The project was first inspired by a simple but common situation in foreign language learning classes: the lack of digital resources that can automatically reduce complex original Arabic texts to simpler versions without altering their meaning. This prevents learner autonomy and complicates teachers' use of adapted material.

b. Educational and scientific purposes

To meet this requirement, the project was structured in four fundamental objectives, each corresponding to a project life cycle phase and an instrumental deliverable:

- Annotation and constitution of an educational corpus:
  - o Gather authentic learning texts from teaching locations (GLOSS[1], Al Jazeera Learning[2]).
  - o Annotate the corpus with readability metrics and linguistic markers in order to make it an exploitable foundation for NLP models.
- Expert lexical resource development:
  - o Develop a vocabulary of basic words with easy levels texts (1 and 1+ according to the ILR annotations (Clark & Clifford, 1988)).
  - o Leverage available databases like Arabic WordNet (Black et al., 2006) and hand-compiled dictionaries.
- Developing a hybrid simplification model:
  - o Combine rule-based techniques (grammar, morphology) with machine learning algorithms (Nassiri et al., 2018a, 2022b, 2023).
  - o Leverage vector representations (CBOW, Skip-Gram) (Preethi Krishna & Sharada, 2020) and pre-trained language models like AraBERT (Antoun et al., 2020).
- Readability and semantic preservation evaluation of simplification:
  - o Quantitative improvement markers: difficulty level categorization (Berrichi et al., 2023; Nassiri et al., 2022a) and an assessment of semantic similarity (Zhang et al., 2019).
  - o Pedagogical impact measures on the basis of accessibility and intelligibility.

c. Strategic alignment

The project is aligned with:

- National priorities in teaching Arabic as a second language,
- Applied research objectives in automatic NLP,
- And the institutional initiative in using AI for inclusive education.

3. RELATED WORKS

In order to situate our contribution within the existing body of knowledge, this section reviews prior work in two complementary domains: (a) automatic text simplification, with a focus on Arabic language processing and hybrid approaches, and (b) project management frameworks relevant to the development of educational technologies integrating Artificial Intelligence.

a. Automatic Text Simplification

Automatic text simplification has been of major concern over the last two decades as a means of improving the accessibility of text to different groups of audience, including language learners, reading disabled individuals, and individuals with low literacy skills. Automatic text simplification research has typically focused on three key levels of simplification: lexical (Paetzold & Specia, 2017), syntactic, and semantic (Al-Thanyyan & Azmi, 2021).

Earlier work on lexical simplification drew largely on rule-based models and linguistic resources. WordNet-based approaches (Bott et al., 2012; De Belder & Moens, 2010) enable synonym identification and ranking according to usage of frequency or lexical length. While such techniques, being understandable and pedagogically appropriate, were in most cases limited in scope and not context-adaptable.

The emergence of parallel corpora (e.g., Wikipedia/Simple Wikipedia[3]) enabled data-driven simplification, training models to identify challenging words and generate simpler equivalent words based on aligned sentences (Biran et al., 2011), (Horn et al., 2014). Deep learning approaches later introduced context-aware strategies like LSBert (Qiang et al., 2021), which leverage the masked language modeling capabilities of BERT (Tenney et al., 2019) to generate contextually appropriate simplifications. These methods have advanced lexical simplification in English and other highly resourced languages significantly.

Compared to others, Arabic received relatively less focus in automatic text simplification. The language is difficult for a variety of reasons such as its complicated morphology, diglossia, and non-availability of annotated corpora. Out of very few early papers in Arabic simplification, Al Khalil et al. (2017) (Al Khalil et al., 2017) introduced one of them for simplifying literary texts of school learners manually. Subsequent alternatives are Khallaf and Sharoff (2022) (Khallaf & Sharoff, 2022), who proposed sentence-simplification models with the use of Arabic-BERT and FastText (Athiwaratkun et al., 2018), and Al-Thanyyan and Azmi (2023) (Al-Thanyyan & Azmi, 2023), who proposed a hybrid simplification system that combined machine translation and transformer-based lexical substitution.

A significant work in the Arabic domain is the release of SAMER (Alhafni et al., 2024) (Alhafni et al., 2024), a large parallel corpus manually simplified for school-age learners. It offers document-level and word-level readability annotations and supports building and evaluating simplification systems for Arabic.

In (Baazeem et al., 2025), the authors presents a novel corpus combining eye-reading measures and textual features, annotated according to subjective readability. This resource represents a significant advance for the empirical analysis of readability in Arabic, promising for NLP and cognition research.

The evaluation of readability in Arabic remains limited by the lack of adequate corpora. (Elmadani et al., 2025) introduced BAREC, a large-scale, finely annotated Arabic readability corpus designed to support research and applications in NLP.

While recent systematic reviews of AI in education underline its transformative impact across K-12 and higher education contexts (*A systematic review of literature reviews on artificial intelligence in education (AIED): a roadmap to a future research agenda | Smart Learning Environments*, s. d.), other surveys in Educational Data Mining highlight the growing role of AI for analyzing learner performance and ensuring ethical use of educational data (*Educational data mining: a 10-year review | Discover Computing*, s. d.).

Despite these advancements, not many systems address the educational use-case of simplifying native Arabic material for second language learners via a systematic and pedagogically guided process. The present work fills this gap by integrating specially designed lexical resources with modern deep learning techniques in a hybrid architecture, and embedding the simplification process along with readability assessment and pedagogical impact analysis.

b. Project Management Frameworks for Educational Technology

Effectively managing projects involving Artificial Intelligence in educational technologies involves more than technical capability; it also requires a formal process which provides assurance that educational goals are connected to the development of technology and the potential users and stakeholders involved. In this situation, established project management models, especially those designed for educational contexts, offers likely a structure for approaching the design/development/evaluation approach in an organized and responsive manner.

In the area of instructional design, one of the most recognizable models of design is ADDIE (Branch, 2009), which stands for Analysis, Design, Development, Implementation and Evaluation. This model works well for educational technology projects, because it emphasizes formal planning, feedback cycles, and thereafter iterative enhancements. In regards to an AI-supported education tool, ADDIE reinforces the importance of interweaving pedagogical intent with tool functionality. For instance, in the Analysis phase, the context learner needs have been established; in Design, the context learner needs are translated into the functions of the system; and the final Evaluation phase allows the team to construct a detailed refinement plan based on feedback.

Another formal model starting to grow in EdTech, is the Agile methodology, and in particular the adaptations for educational innovation, such as EduScrum (Wijnands & Stolze, 2019), or Agile-LMS models. Both support rapid-prototyping, flexible collaborative teams, and work in progress, all critical to the emergence of projects that will be utilizing evolving AI tools and unknown user needs. Agile's iterative sprints correspond nicely with the exploratory nature of NLP integration into education by allowing for constant experimentation with continuous incorporation of feedback from educators, learners, and linguist professionals, as relevant.

Also, models like PMBOK (Project Management Body of Knowledge) (Project Management Institute, 2008) have been referenced to assist with resource management, stakeholder roles and responsibilities, risk management, and timeline management at a broader level. Although PMBOK does not specific to education, its structured approach aligns with education frameworks, and aids decision-making, especially in relation to uncertainty - a common factor with AI and EdTech.

In our project to simplify Arabic educational texts through a blended AI approach, we adopted a blended management strategy. The earlier phases followed the ADDIE model, particularly the analysis of learner profiles and determining educational needs. The latter phases adopted Agile characteristics, so we could gradually develop modules, iterate the linguistic simplification techniques, and integrate real-time feedback by educators. Blended strategies provided us pedagogical validity, technical credibility, and adaptation throughout the entire horizon of the project.

Overall, our project was not only professional but anchored in project management systems. We organized our methodology, with the process encumbered through comparisons of project management standards, while responding to the calls of the creative agency required by AI in education.

4. METHODOLOGY AND PROJECT MANAGEMENT STRATEGY

The process needed to develop an Arabic text simplification system for education was to have a structured approach that brought together linguistic knowledge, AI technologies and project management strategies. The project had a hybrid methodology using both rule-based and data-driven approaches and progressed through established stages of educational project management that include planning, action, review and evaluation.

a. Project Planning and Team Structure

The project was initiated with a detailed planning process that established measurable outcomes, the targeted user group (Arabic as a second language learners) and the linguistic challenges (e.g. lexical complexity, syntactic complexity) to be addressed. Expected results were an operational text simplification module in educational platforms.

Roles and responsibilities were distributed:

- Linguists were to curate and annotate educational texts,
- Natural language processing engineers to build and train the simplification models,
- Project coordinators to administer timelines, deliverables and research ethics.

b. Hybrid Linguistic–Computational Framework

To achieve the right trade-off between educational interpretability and computational efficiency, a hybrid architecture with both symbolic (rule-based) and statistical (data-driven) components was adopted in the project:

- **Knowledge-based resources:**
  - Arabic WordNet and synonym lexicons,
  - Frequency-based lists of vocabulary congruent with school levels,
  - Manually curated and annotated corpora for training and evaluation.
- **Data-driven models:**
  - Distributional semantics (CBOW, Skip-Gram),
  - Transformer-based models fine-tuned for Arabic (AraBERT),
  - Evaluation metrics for semantic similarity and fluency (BERTScore).

With this integrated approach, the system augmented its linguistic simplifications with knowledge-based and context-appropriate pedagogical changes by employing differing degrees of knowledge-based and semantic embeddings.

c. Workflow Design

The simplification system consisted of a modular and iterative workflow designed specifically to the educational NLP context. Each of the components was designed with the intention of traceability, reproducibility, and to align with pedagogical outcomes. The workflow included the following stages, which are interrelated:

1. **Corpus Acquisition and Preprocessing:** Educational texts were compiled initially from L2 platforms (GLOSS, Al Jazeera Learning). After sourcing, the texts were segmented, cleaned, and filtered for linguistic quality and educational relevance.
2. **Annotation Scheme:** The educational texts were annotated with readability levels (Easy, Medium, Hard) based on their ILR scale.
3. **Lexicon Development:** A lexicon of "easy" Arabic words was constructed using original beginner-level Arabic texts and relevant frequency analysis. This lexicon served as a guide for the complex word identification step.
4. **Simplification Pipeline:** A modular simplification pipeline is at the heart of the system designed to convert educational texts with complex Arabic into an easier, more learner-friendly form while preserving meaning. The pipeline is a blend of rule-based and machine learning, and the following components are part of the pipeline:
   i. Identification of Complex Words (ICW): This step consists in identifying complex words in the text which are difficult to understand.
   ii. Substitute Generation (SG): After identifying a complex word, we create a list of simpler synonyms that could be used as substitutes.
   iii. Substitute Selection (SS): The synonyms generated above will be ranked from simple to complex and the simplest synonym will be chosen to substitute for the complex word in the text.
5. **Evaluation Stage**: Simplified outputs were evaluated using readability classification models (before/after) and semantic similarity (BERTScore).

As an example of the potential impact of our system for lexical simplification, consider the following real classroom scenario: middle-school students in an astronomy unit will encounter terms such as " " (galaxy) or " " (constellation). Our approach will provide synonyms or brief sales or definitions- for " ", we will provide " " ("a large group of stars"), which allows students to focus on the science rather than the vocabulary. And for students who struggle with reading, like students with dyslexia, the same functions will apply as complex type words such as " " (empire) will be synonymous with a simpler word or a brief definition such as " " ("a large state ruled by a king"). Every student will have equitable access to learning.

The system is also benefical in multilingual classrooms where a student might encounter a culturally specific term that may not click for a non-native Arabic speaker. For example, the Arabic term " " or "poetic contest" can simply be defined as " " or "competition in reciting poetry", giving students meaning and the contextual basis. Accordingly this system can help a teacher or an assessor with feedback and assessment. For example, where an assignment indicates reading, a teacher is trying to assess difficult words that may be the source of repeated mistakes, say, "   " or "conclusion", and can highlight the difficulty upon reading, and provide a simplified definition ("   " – "result") in efforts to execute teaching and reinforce learning.

The integration of a technology system into digital or online teaching and learning also creates opportunities for greater accessibility. For example, a student facing a word, "   " which translates as "innovation" could instantly make sense of a simplified word to support their engagement in the reading ("   " – "a new idea"), allowing for comprehension in a digital or e-learning scenario. These scenarios represent how the system put verbs, nouns and adjectives into action through lexical simplification that can occur in an authentic educational setting and supports inclusivity moving forward, by providing a best effort to meet the goals of Universal Design for Learning (UDL).

d. Agile Iterations and Feedback Loops

The project employed an agile model with a focus on short sprints. This allowed for a better sense of progression through repeated short iterations of focused work. Each sprint aimed for a functional piece (e.g. simplified rules, improved model, evaluating tools) but also considering pedagogical objectives.

At the end of each sprint, we led retrospective meetings to determine ways to improve, learn from mistakes, and pave the way for future development. The design made iterative work user-led, adjustable, and pedagogically sound.

Along with feedback from educators and linguists, feedback loops from learners are also needed in the evaluation phase. The learners are asked to engage with the simplified texts and evaluate on a scale how clear, easy to understand, and useful they felt it was. Their feedback then informed changes to be made to the substitution strategies and interface design. This approach resulted in a system that will be not only linguistically sound, but one that also considered learners' cognitive and contextual needs, moving educational experiences to a more inclusive and personalized model.

e. Risk Management

A key component of the methodology was the identification of risk, such as:

- Data bias with educational corpora,
- Overfitting due to a small number of annotated samples,
- Regions (or features) of simplified texts produced by the automated process not representing the pedagogical intent of the project.

The project had built-in strategies to manage risks through regular project review sessions, using double annotation for critical datasets, and conducting teacher validation of simplified materials identified as pedagogically suitable prior to classroom integration. The project also adhered to ethical principles of conducting research, such as ensuring appropriate data sourcing, guaranteeing suitable learner representation aspects, and transparency of the AI models applied for simplified text instances within the corpus.

The methodology described above drew on academic rigor by integrating linguistically-rich annotation processes, state-of-the-art NLP processes, and iterative project management techniques to develop an Arabic text simplification system that could be introduced and used within the education context with relative ease. The hybrid form and nature of the approach achieved an effective integration of AI, and careful scrutiny of the simplifications produced against representation of educational purposes, to obtain an effective and scalable, learner-centred technology that aligns with pedagogical principles and purposes.

To provide a more structured view of potential threats to the project and their mitigation, we provide a risk register summarizing key risks, their likelihood, potential impact, and mitigation strategies:

Table 2. Risk register of the project

Risk	Likelihood	Impact	Mitigation Strategy
Low teacher engagement	Medium	High	Early involvement in co-design sessions and regular feedback cycles
System performance failures	Low	High	Cloud-based deployment with automated monitoring and backup systems
Data privacy and usage concerns	Low	High	Adherence to ethical guidelines and data protection regulations
Model obsolescence due to curriculum updates	Medium	Medium	Yearly model updates and retraining aligned with curriculum evolutions
Institutional reluctance to adopt tool	Medium	High	Pilot testing, reporting of learning outcomes, and stakeholder workshops

5. CORPUS CONSTRUCTION AND ANNOTATION

The quality and representativeness of any NLP project are the single most important aspect. In the project of educational Arabic text simplification, it was essential to create a pedagogically meaningful, linguistically diverse corpus so we could produce simplifications that were accurate and contextually aware.

a. Source Selection and Pedagogical Relevance

The corpus was created with a priority of Arabic L2 learning. Two sources were selected as they matched our goals of language instruction:

- GLOSS (Global Language Online Support System) – Graded Arabic texts in cultural, scientific, and current affairs topics. We have created two corpora from this source: GLOSS-Reading (GR) and GLOSS-Listening (GL).
- Al Jazeera Learning Portal – Offers structured Arabic L2 learning material for classroom use with authentic materials based on media.

The corpus aimed to represent a range of linguistic difficulty levels, from beginner to intermediate, to reflect the progression of L2 learners and support adaptive simplification strategies.

b. Corpus Size and Structure

The final dataset comprised of 819 educational Arabic texts, consequently with an average of 120-250 words each. Each document was divided into sentences; in its total the dataset contained 14,032 annotated sentences or about 150,000 words.

Table 3 constitutes the distribution of the corpus and is classified into 5 levels of difficulty:

Table 3. Distribution of Corpora over difficulty levels

Level	GLOSS-Reading	GLOSS-Listening	Al Jazeera Learning	Total
1	34	43	12	89
1+	29	18	220	267
2	95	65	54	214
2+	68	62	27	157
3	45	39	8	92
Total	271	227	321	819

c. Dimensions and Guidelines for Annotation

To satisfy the dual purposes of simplification and educational validity, each of the texts were annotated in three ways:

- Readability level: each text was assigned a readability level based on ILR/ ACTFL (Clark & Clifford, 1988) descriptions of levels of ease:
    - o Easy (ILR 0-1+),
    - o Medium (ILR 2–2+),
    - o Hard (ILR 3 and above).
- Linguistic features: each text was annotated for average word length, morphological difficulty, level of rare vocab, ...
- Educational validity: each sentence was assigned a usage context marker (explicit reading comprehension, vocabulary development, etc.).

d. Tooling and Automation Support

The annotation procedure was semi-automatic using a set of NLP tools geared towards Arabic:

- MADAMIRA (Pasha et al., 2014) was the tool leveraged for morphological analysis and part-of-speech tagging,
- AraNLP and the CAMeL Tools (Althobaiti et al., 2014) assisted with preprocessing (e.g., machine learning pre-training) and named entity recognition,
- And custom Python scripts assisted for integration of readability rules and scored using a frequency-based method.

In depth profiling of the corpus also involved computing distributions of key grammatical categories (e.g., nouns, verbs, adjectives, adverbs, and function words) across all levels of difficulty. This provided pertinent data for establishing lexical imbalances, and designing the targeted set of simplification rules for verb-rich texts or texts that contain many descriptors.

Table 4 indicates percentages of all grammatical categories for each level of difficulty in the corpus.

Table 4. Distribution of grammatical categories by difficulty level

Level	Nouns (%)	Verbs (%)	Adjectives (%)	Adverbs (%)	Tools (%)
1	43.01	11.37	24.62	3.34	17.66
1+	40.28	12.52	25.12	3.15	18.93
2	37.09	13.72	25.98	20.12	20.12
2+	36.18	13.63	24.37	3.27	22.55
3	34.18	14.71	23.94	3.44	23.73

It is important to note that the grammatical category "Nouns" includes both common and proper nouns, and "Tools" is a catch-all category for any remaining tags that are not nouns, verbs, adjectives, or adverbs like prepositions, conjunctions, and particles. This grammatical distribution showcases the manner in which textual complexity grows as your level increases, and will support you in creating both the rule-based, and data-driven simplification elements for students to engage with.

e. Lexical Simplification Tags

As well as sentence level annotation, the corpus was enhanced with word level annotation to support lexical simplification. We specifically annotated:

- **Complex words** based on learner difficulty and frequency,
- **Simpler synonyms**, at the appropriate educational level,
- **Semantic equivalence markers** that indicate meaning can be preserved while simplifying terms.

For this project, we created a custom elementary lexicon using vocabulary consistent with the linguistic content found in a corpus of 356 easy-text texts. The lexicon has 13,065 elementary words with semantic and lexical content targeted to reflect the vocabulary available to Arabic L2 learners at the novice level.

After constructing this list of simple words, we evaluated text at higher levels (levels 2, 2+, and 3) to determine the percentage of words outside the elementary lexicon. The results, shown in Table 5, detail the percentage of complex words.

Table 5. Percentage of words in difficult texts that do not appear in the simple word lexicon

Level	% of words not appearing in the lexicon of simple words
2	74.80%
2+	75.55%
3	72.02%

In our findings, we can note that roughly three-quarters of the words in the difficult level do not belong to the list of simple words, and therefore are candidates to be counted as complex words. This very specific lexical tagging provides the corpus-based simplification module with a supervised learning framework, such that there are contextually-valid mappings between complex words and their simpler synonyms in a manner that both maintained semantic fidelity, as well as pedagogical validity.

6. HYBRID SIMPLIFICATION ARCHITECTURE

The key innovation in this project is a hybrid lexical simplification system that is appropriate for Arabic education texts, and uses a combination of rule-based linguistic knowledge and deep learning models to produce simplifications that are semantically, contextually and pedagogically appropriate.

a. System Overview

This simplification pipeline consists of the following main steps:

- Complex Word Identification (CWI),
- Substitutes generation (SG),
- Substitutes selection (SS),

In each step we combine both knowledge and ML insights to enhance our outputs and instill trust from users.

Figure 1 depicts the steps in our simplification process which takes a raw text that is considered difficult input and attempts to the output of a simplified text.

Figure 1. Process adopted to simplify an Arabic text

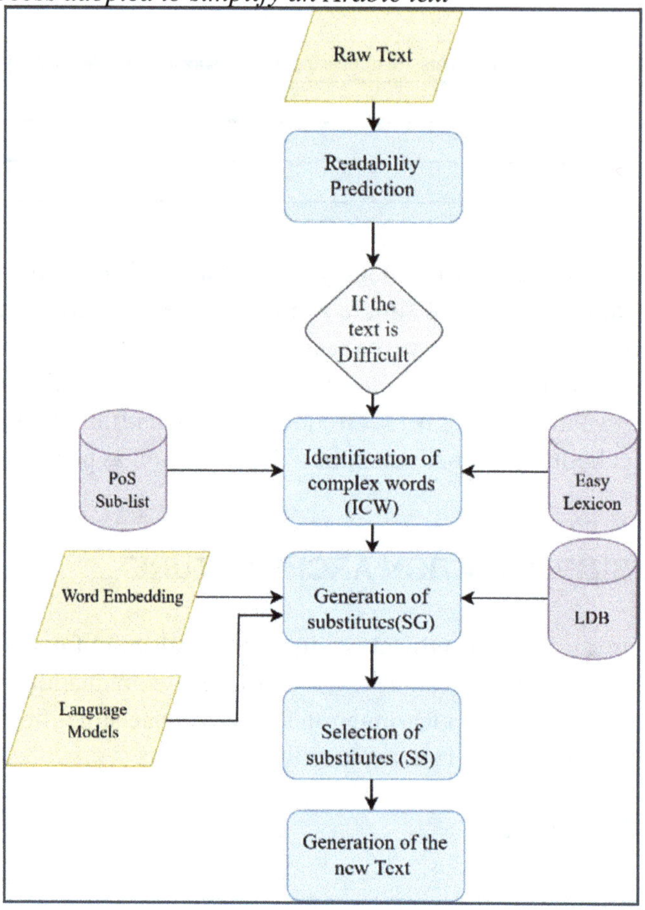

b. Complex Word Identification

An important problem for managing Arabic text simplification in the context of educational applications is consistently and reliably identifying words that may be complex for L2 learners. In our hybrid AI approach, we tackle this complex word identification problem in two steps through a multi-criteria filtering method depending on linguistic rule-based logic and data-informed heuristics. This methodological approach will hopefully provide a pedagogically appropriate balance with computational precision.

□ Step 1: Filtering Based on Grammatical Categories

Using the methodology that Al-Subaihin and Al-Khalifa (2011) (Al-Subaihin & Al-Khalifa, 2011) proposed, we began our simplification pipeline by filtering the words in our corpus of documents by part-of-speech (PoS) tags from the MADAMIRA morphological analyzer. We were particularly interested in the grammatical categories associated with more complex lexical difficulty for L2 learners. The categories we focused on were nouns, verbs, adjectives, and adverbs, and we excluded proper nouns, articles, and prepositions, which are categories that were generally considered basic or accessible for learners and did not need to be simplified.

After applying this PoS-based filter to a corpus of 463 documents from levels 2, 2+, and 3, we started with 155,835 words (punctuation, numbers removed) and reduced the potential candidates to 112,956 words (72%) that were appropriate for complexity evaluation.

□ Step 2: Filtering Based on a Custom Simple Word Lexicon

Next, we leveraged the handcrafted simple word lexicon, which compiles elementary vocabulary drawn from easy-level texts (levels 1 and 1+). We assume that words present in this lexicon are already known to learners at earlier stages and do not require simplification. Filtering the 112,956 candidate words against this lexicon further reduced the pool to 45,272 words (29% of the original total). These remaining words are deemed complex and represent the primary targets for lexical simplification.

The next stage of the simplification pipeline requires complex words, when identified in the corpus, to be replaced with simple, pedagogically appropriate synonyms. The whole substitution process is a prominent facet of the system, since it impacts the intelligibility of the simplified output, as well as faithfulness to the original text. The Substitution Generation (SG) module uses three complementary approaches: Substitution via Linguistic Databases; Substitution via Word Embeddings; and Substitution via Pretrained Language Models.

c. Generation of substitutes via Linguistic Databases

Since this phase required a place to begin, we developed an SG module, which generates and ranks synonymous candidates for each complex word found in the original text, based on relation to context, and simplicity.

In this step, we used structured lexical knowledge to yield simpler synonyms with lexical resources that were built by expertise. We used two principal linguistic databases:

- **Arabic WordNet (AWN)** (Black et al., 2006): A structured lexical ontology with 11,269 lemmatized syn sets classified by grammatical categories.
- **ArabicLexicon (AL)**: A lexicon built from several Arabic dictionaries, and contains 16,832 lexical entries, plus their synonyms.

Given that both linguistic resources, Arabic WordNet (AWN) and ArabicLexicon (AL) keep entries in a form of a lemma, we need a structured transformation process to generate accurate and grammatically correct synonyms for inflected complex words. The checklist will ensure we maintain morphosyntactic integrity and semantic equivalence for granting substitutions. The synonym generation process involves the following distinct stages:

a. Morphological Analysis: The first step involves the morphological analysis of the complex words with MADAMIRA. MADAMIRA extracts a number of linguistic parameters such as lemma, morphological features (gender, number, tense, mood, case, and voice), and any clitics (eg, prefixes, suffixes, pronouns, particles) that can be attached.

b. Synonym Collection from Lexical Databases: After having established the lemma for the complex word, it is then compared against the entries in AWN and AL to extract a list of semantically equivalent lemmas. These candidates for the synonym pool are now the basis for the semantic synonym substitute word options, and are categorized by part of speech to ensure that they function grammatically the same way, as implied by the original word.

c. Morphological Generation: For each synonym lemma, the CAMeL Tools morphological generator yielded the respective inflected form which was morphologically the same as the original word. Depending on the original lemma's grammatical features, the inflection preserved the same gender, number, case (for nouns), and tense, mood, voice, and person (for verbs) properties. For example, if the original word had been a feminine plural noun in the accusative case, the generated synonym would also be designated the same grammatical form as the original word. For verbs, the inflection preserved tense, aspect, mood, voice, and person.

d. Clitic Reconstruction: Finally, clitics that were originally on the complex word-like definite article ( ), coordinating particles (,) or pronominal suffixes- are reattached to the inflected synonym. This step ensures that the replacement word fits exactly with the syntactic and discourse context of the original sentence.

d. Generation of substitutes via Word Embeddings

To address the limitations of traditional lexical resources, we implemented a second method based on word embeddings (WE), which allow semantic similarity computation through high-dimensional vector representations of words. Word embeddings are grounded in the distributional hypothesis—that words appearing in similar contexts tend to have similar meanings. Trained on large-scale Arabic corpora, these models learn to capture semantic relationships between words through unsupervised learning. In our framework, we integrated two widely adopted architectures to model these semantic similarities:

a. **Continuous Bag-of-Words (CBOW)**: Predicts a word based on its surrounding context within a fixed-size window (Mikolov et al., 2013) (Mikolov et al., 2013).
b. **Skip-Gram**: Predicts surrounding context words given a target word, better capturing low-frequency words and flexible contextual associations (Mikolov et al., 2013) (Mikolov et al., 2013).

Each complex word was projected into the embedding space, and its cosine similarity was calculated against all other word vectors. The five most similar words were automatically selected as candidate substitutions. This data-driven selection process was then reviewed through manual evaluation by the project team, confirming that the nearest neighbors frequently included semantically coherent and pedagogically simpler alternatives. This step ensured that the substitutions aligned not only with computational similarity, but also with the educational goals of improved accessibility for Arabic L2 learners.

e. Generation of substitutes via Pretrained Language Models

The major innovation in our approach lies in leveraging AraBERT's MLM capabilities within a context-aware substitution framework tailored for Arabic text simplification. Unlike the simpler lexical substitution methods, this method predicts possible replacements since it takes into account both left and right contexts of the masked complex word, allowing for more natural and semantically accurate simplifications.

The first stage begins by creating a copy of the original sentence and replacing the complex word in the copy with the special token [MASK]. This provided pair of sentence-type input, bracketed with the standard [CLS] and [SEP] tokens, is input into AraBERT which returns a vocabulary probability distribution for the masked position.

The model will suggest candidates substitutes that are ranked according to likelihood from this distribution. To ensure these candidates are not only grammatically plausible but also semantically faithful to the original word, we also filtered candidates using contextual similarity, such as BERTScore.

Figure 2 shows an example of how the model works. In fact, if we look for potential synonyms for the word " " in the sentence " ", we apply the MLM task to the concatenation of this sentence (the beginning of the sentence is identified by the general symbol "[CLS]" and its end by the general symbol "[SEP]") with the same sentence where we have replaced the word " " with the symbol "[MASK]". The model thus proposes a list of substitutes " ", " ", " ", " " et " " with their probabilities. We note that some candidates, for example " ", show a strong correspondence with the complex word " ".

Figure 2. Example of AraBert substitute generation

For our language model-based method, we used the AraBert model available on the "Hugging face transformers"[4] platform. This model suggests synonyms for words we want to simplify. The model was pre-trained on 70 million Arabic sentences, representing a configuration of 24 GB of text. It includes 12 encoder blocks, 768 hidden dimensions, 12 attention heads, a maximum sequence length of 512 tokens, and a total of 110 million parameters

f. Selection of substitutes

After generating a list of potential synonyms, the system must make a choice for the one that is the single-best educational substitute. In the hybrid model, this is done in two steps, by applying linguistic filtering rules and probability- or frequency-based ranking.

a. Filtering the Candidate List

If the SG step provides an empty list, the complex word is unchanged. If candidates are generated, there are two filtering steps that happen:

- Lemma Duplication Filter: Any candidates with the same lemma as the original complex word are removed. This ensures that we are not proposing simple inflectional forms instead of real true synonyms.
- Complexity Filter: To ensure that we only keep learner-friendly words:
  - o Any candidates not found in the Elementary Lexicon are discarded unless found in the High Frequency Dictionary for Arabic (Nassiri et al., 2018b).
  - o The rationale is words that we encounter frequently in natural language corpus, or already viewed as elementary, are more likely to be understood by L2 learners.

If after the filtering step around the candidates related to a particular complex word still has members, the model will move to the step of selecting the best member substitution. Otherwise, the word will not be replaced in the simplified sentence.

b. Ranking and Selecting the Optimal Substitute: the surviving candidates are ranked according to the generation method:
   - Embedding or AraBERT Based Generation
     - o Candidates are ranked on their model probability (WE cosine similarity, MLM (AraBERT probability)).
     - o Rule (a): If only one candidate appears in the Elementary Lexicon, then that one is selected. If multiple appear in the Elementary Lexicon, the candidate with the highest model probability is selected.
     - o Rule (b): If none appears, the same reasoning was used to rank them for the High Frequency Dictionary.
   - Linguistic Database Generation: The rules (a) and (b) are applied in the same way, but ranked by corpus frequency instead of model probability.

With a built-in dual filtering and ranking approach, you can rest assured that your final-substitute is contextually appropriate and pedagogically easier, and completely in-sync with the "content-level" objectives of Managing Educational NLP Projects with Hybrid AI.

7. EVALUATION AND RESULTS

To evaluate the potential effectiveness of the proposed hybrid simplification system, a more global evaluation approach was employed to target technical effectiveness and pedagogical possibilities. The two-part evaluation focused on two key ascribed criteria for educational text simplification: enhancement of readability and preservation of semantics.

a. Evaluation Framework

In order to assess the potential effectiveness and educational relevance of our proposed Arabic text simplification system, we developed a multi-dimensional evaluative model which included evaluation criteria designed to satisfy the dual criteria of linguistic quality and pedagogical value. The evaluation model was focused on three primary research questions:

- **Q1 – Readability Gain:** To what extent did the system improve the readability of educational Arabic texts for L2 learners?
- **Q2 – Semantic Fidelity:** Did the output correctly represent the meaning and intention of the source text?
- **Q3 – Pedagogical Acceptability:** Did Arabic language instructors find the simplifications to be useful, reasonable, and valuable educationally?

To answer the research questions in a comprehensive manner, we used both automatic metrics and human evaluations, which provide a balanced and comprehensive modality of evaluation.

b. Readability Assessment

To determine if the simplification process was successful in reducing the input text complexity, we constructed a supervised readability classifier that was trained on the annotated Arabic corpus exposed in previous sections. The readability classifier was used to confirm if the simplified texts were assigned a lower difficulty level (i.e., reduced from Level 3 to 2+ or 2).

This automatic feature will provide a quantification for a simplification rate based on the percentage of the texts that were simplified to a lower difficulty level. Each of our experiments corresponding to one of the three substitution strategies stated above: (1) substitution based on a linguistic database (LDB), (2) substitution based on distributional word embeddings (WE), or (3) substitution based on contextual language models (LM), followed the same hybrid models that used combined resources.

Table 6. Simplification rates by substitution strategy

Substitution Method	Level 2	Level 2+	Level 3	Overall Simplification Rate
Linguistic Database (LDB)	5.60%	7.64%	6.52%	6.47%
CBOW (WE)	7.94%	8.20%	11.95%	8.85%
Skip-Gram (WE)	9.34%	10.82%	14.13%	10.79%
CBOW + Skip-Gram	10.28%	17.83%	30.43%	16.84%
AraBERT (LM)	9.34%	27.38%	35.86%	**20.30%**
Hybrid (LDB + WE)	10.28%	12.73%	21.73%	13.30%
Hybrid (LDB + LM)	20.09%	15.28%	20.65%	**18.57%**

The assessment of the three main substitution strategies, highlighted different performance characteristics that guide the modeling of Arabic text simplification systems.

LDB models had the lowest simplification rate of all models. In fact, the model was able to achieve a 6.47% simplification rate. While WE based models had a significant gain when compared to linguistic database models. Both WE models were able to achieve significant simplification rates. The CBOW model achieved an 8.85% simplification rate, while Skip-Gram was able to perform better with a 10.79% simplification rate. This performance difference is likely the result of Skip-Gram better handling rare or poorly represented word contexts. Most importantly, when the CBOW and Skip-Gram model is combined (CBOW + Skip), the simplification percentage nearly doubled to a 16.84% percent simplification rate. This highlights the benefit of adding together multiple semantic spaces when enriching the number of synonyms through a vector space embedding. Additionally, word embeddings are scalable and provide larger coverage when comparing to linguistic databases.

The contextual language model, AraBERT, was the best performing of all individual models with a simplification rate of 20.30% due to the model's ability to understand both left and right context through masked language modeling, this allowed the model to suggest semantic propositions and make correct grammatical substitutions. This modeling flexibility becomes more important when dealing with Arabic because of how much meaning depends on morphological and syntactic structure.

While evaluating manually simplified texts, we found that the suggested synonyms from the linguistic databases were more frequently used and easy to comprehend. The shift in ease of comprehension inspired us to pursue a hybrid model that incorporated the linguistic databases with neural synonym generation. In our first experiment we sought to find synonyms using the combined WE model (CBOW+Skip-Gram) for complex words not present in any of the two LDB. This hybrid model increased our simplification rate from 6.47% to 13.30%. In our second experiment rather than WEs we used LMs, which increased the simplification rate from 6.47% to 18.57%.

To summarize, although linguistic resources yield a useful base structure and provide some amount of linguistic reliability, they are inadequate when deployed exclusively for large-scale context-aware simplification. On balance, they suffer from a limited set of pedagogically appropriate synonyms and are limited by requiring lemmatized entries, which are not created with inflected forms or nuanced contexts. Neural models, and particularly contextual models such as AraBERT, are much better functional means to identify substitutions that meet their semantic intent and syntax satisfaction. While hybrid models are best middle ground, they ensure that simplifications meet both educationally soundness and linguistic correctness which are useful for the purpose of developing Arabic educational NLP project using a Hybrid AI architecture.

c. Performance Evaluation

In order to clarify how effective these models are for this simplification process, we evaluated their performance for identifying and substituting complex words through a statistical review of their assessments, which is outlined in Table 7 and incorporates three benchmarks:

- **ICW**: the percentage of words identified as complex in the input corpus,
- **SG**: the percentage of those complex words for which a substitution was generated,
- **SS**: the percentage of complex words that received at least one **simple** substitute (according to the Elementary Lexicon or frequency-based filtering).

Table 7. Statistics on complex words and their substitutes provided by models based on linguistic databases

Substitution Method	ICW (%)	SG (%)	SS (%)
Linguistic Databases (AWN + AL)	24,91	10,45	7,59
CBOW (WE)	24,89	93,42	54,39
Skip-Gram (WE)	24,89	94,74	58,31
AraBERT (Lang. Model)	24,89	100	88,85

Our analysis revealed that around 25% of words in difficult texts were identified as complex by all these methods. However, only a small fraction of these complex words had synonyms that could be used for simplification. In the Linguistic Databases, synonyms were found for at most 10.45% of complex words. Hence, only 7% of complex words will be substituted by a simple alternative. This suggests that the low simplification rates are mainly due to the limited number of synonyms available in these databases. For word embedding methods, we notice that The CBOW and SkipGram models were able to suggest substitutes for around 93% of the complex words. The remaining 7% lack potential substitutes, as they are categorized as out-of-vocabulary words within the Word embedding model. Both models were only able to simplify about half of the complex words. Finally, the AraBert contextualized model, manages to generate substitutes for all complex words, and in at least 88% of cases proposes a simple substitute.

Whereas simplification rate alone provides one quantitative measure of success, it does not adequately reflect the new level of change introduced by the system. In educational contexts, it is also necessary to assess how much easier a text has been made, especially in terms of its proper match with the students' levels of proficiency. To investigate this, we measured the degree of simplification that each model achieved, and we scored the outcomes based upon whether the simplified version was:

- One level easier than the original (e.g., Level 3 → 2+),
- Or two levels easier (e.g., Level 3 → 2 or 1+).

Table 8 shows the percentage of simplified texts that were moved down from one or two readability scores for all methods. The results are clear. Most of the simplified texts, more than 80%, moved to the first lower level of the original readability category. This reinforces the system's more conservative and effective approach to simplification, which retains meaning and grammatical accuracy while reducing complexity in a pedagogically appropriate way.

Table 8. Results regarding the degree of text simplification

Method	Reduction by one level	Reduction by two levels
Linguistic Databases	86.6%	13.4%
Word Embedding	80.77%	19.23%
AraBert	84.05%	15.95%
Hybrid (LDB + WE)	83.88%	16.12%
Hybrid (LDB + AraBert)	82.56%	17.44%

d. Semantic Similarity and Meaning Preservation

In educational text simplification, the importance of conserving original meaning is as valuable as improving the readability of the text. Simplified texts are expected to be faithful representations of the source text so that learners are not misinformed or taken away from the essential information. To quantitatively measure semantic similarity, we utilized BERTScore [7], a prevalent metric that assesses the similarity of two (or more) sentences based on contextualized word embeddings from pre-trained language models.

BERTScore is distinct from BLEU and ROUGE because BERTScore does not utilize a purely surface-level state. While BLEU and ROUGE assess a simplified text's surface similarity to a reference text, BERTScore assesses the semantic similarity by looking at each word in the simplified text with the closest context-appropriate counterpart in the original text using its pre-trained transformer model. In other words, BERTScore metrics can factor in lexical and syntactic variance, while maintaining semantic similarity.

To have meaningful and generalizable results, we applied BERTScore metrics to a 10% stratified random sample of the educational texts from each difficulty and substitution method. Each subset was evaluated with AraBERT as the language model which has demonstrated high accuracy capturing semantic nuances in Arabic. The measures come from Table 9, reports the precision, recall, and F1-score, with F1-score as the primary measure of overall semantic variance.

Table 9. BERTScore results with different simplification models

Method	Level	Precision(%)	Recall(%)	F-score(%)
Word Embedding (WE)	**2**	88.52	88.56	88.54
	2+	91.59	**91.69**	91.63
	3	81.22	81.45	81.34
AraBert	**2**	92.4	90.96	91.72
	2+	91.77	99.46	91.14
	3	86.79	84.37	85.65
Hybrid (LDB + WE)	**2**	89.78	90.51	90.12
	2+	89.81	90.60	90.18
	3	84.17	84.19	84.15
Hybrid (LDB + AraBert)	**2**	**95.07**	**94.80**	**94.93**
	2+	**92.49**	90.98	**91.72**
	3	**90.53**	**90.26**	**90.39**

We point out that for levels 2 and 3, the semantic quality of the simplification of the AraBert model is much better than that of the WE model, and these qualities are almost close for level 2+. Moreover, combining these models with the LDB-based model significantly improves the semantic quality of the simplification. The highest F-score was achieved for the least complex texts of level 2 (94.93%), and the lowest for the most complex texts of level 3 (90.39%).

An analysis of both tables (Table 6 and Table 9) reveals that evaluating simplification through readability metrics and semantic similarity with BERTScore shows the AraBert approach combined with the linguistic database to be an effective method for generating and selecting word substitutions.

To better illustrate the effectiveness of the proposed simplification method, we present several output examples generated by the system. These examples showcase how complex lexical items were automatically replaced with simpler alternatives, while maintaining the overall meaning and fluency of the sentence. They reflect the system's ability to produce contextually appropriate simplifications, guided by both semantic similarity and linguistic resources.

Original sentence:

“ ____ ____ ___ .”

Simplified sentence:

“ ____ ____ ___ .”

In this example:

- “ ” (current challenges) → “ ” (current difficulties): simpler synonyms with reduced lexical weight.

e. Error Analysis

A qualitative examination was completed on when error or less-than-optimal outputs occurred. The main issues were:

- Though there could be some occasions where gender/number agreement mismatch could occur after substitutions.
- Difficulty simplifying named entities and idiomatic language.
- Over-simplifying scientific discourse as to render it technically unhelpful.

Recommendations for action:

- Created some post-processing for morphological agreements.
- Improved stop word lists and some term holding rules.
- Created filter with domains in mind (e.g., scientific vs. narrative)

8. DISCUSSION AND EDUCATIONAL IMPLICATIONS

The development and evaluation of a hybrid Arabic text simplification system does not merely present additional challenges in the limited natural language processing literature for under-resourced languages for future research, but new understandings will inform future educational artificial intelligence project design and management. This section will reflect on the implications of the overall research study and outline its implications for inclusive education, curricular adaptation, and sustainable research practice.

a. Enhancing Inclusive Education through NLP

Most notably, the project's contributions revolve around educational equity and access. In many Arabic-speaking contexts, complexity in language creates barriers in learning for:

- Non-native Arabic learners (L2),
- Students with disabilities, learning disabilities, or reading challenges,
- Children or adults in rural and under-served areas.

The proposed system provides real-time text simplification in a scalable way that affords personalized learning spaces and differentiated instruction. These spaces support Universal Design for Learning (UDL) and inclusive pedagogy where learning resources adapt to the learner, rather than the learner adapting to them.

Teacher-evaluated simplified texts also add a layer to text simplification and ensure that outputs are pedagogically valid and usable in classroom contexts, real and virtual.

b. Bridging AI Innovation with Pedagogical Needs

A common issue in education-focused artificial intelligence applications is a disconnect between system functionality, or 'how it works,' and classroom practice. This project accounted for this by including educators at all stages of the process, including:

- In corpus construction and annotation
- In designing lexicons for levels of complexity
- In evaluating the system outputs accounted for this by.

Involving educators in a design process like this gave support to the notion that the system worked algorithmically, addressing issues of 'how to' but, just as importantly, educators' 'expected use' in their instructional practice as language instructors.

This model indicates a human-in-the-loop approach that merges data science with educational evidence, and can serve as a potential reusable blueprint for other NLP-for-education projects.

To promote sustainability across the research phase of this project, the researchers put forward a maintenance and deployment plan. To facilitate planned ongoing maintenance, a maintenance cycle has been proposed involving the regular retraining of the model using new learner data used by the Lead Student, updates to the lexicons, and incorporation of user feedback. For sustainability, the researchers see working with educational settings for possible local hosting and alignment with the curriculum. They are also developing an administrative panel to let administrators manage the system, and the teachers set the levels of simplification and monitor activity.

c. Relevance for Low-Resource Languages and Cross-Lingual Education

Arabic has unique challenges—morphological richness, diglossia, and low-resource languages—but many of the ideas here can generalize to a broader set of low-resource languages with education in mind:

- The hybridization approach (of rules and deep learning) is generalizable,
- The pipeline for corpus building is generalizable and can easily include domain knowledge with the correct data,
- The evaluation which employs both readability and semantic preservation is language independent.

In multilingual education, this work highlights the need for practitioners to adapt AI systems to their own linguistic and cultural context rather than simply adopting an English-centred approach, while at the same time learning about other multilingual, and by default, cross-lingual opportunities for learning.

d. Project Management Insights in AI-Driven Educational Research

When considering some project management best practices, this case exemplifies a reiteration of a number of best management practices:

- a clear statement of pedagogical objectives at the start of the research (and not just technical KPIs),
- phase-divided work (corpus, resources, modelling, evaluation),
- a mixed methods assessment (quantitative + qualitative),
- scalability with modular system design and open lexicon frameworks.

The experience of the project also highlights the important of interdisciplinary coordination - linguists, AI researchers, instructional designers and teachers - when managing educational innovation project as it can become quite complicated.

Additionally, in a study like this- the use of open tools, pre-training models and scalable frameworks promote sustainability and reproducibility, specific to the academic study which often has limited funding.

e. Ethical Considerations

Finally, no ethical consideration would be complete without reminding us of the ethical dimension. Simplification cannot lead to:

- loss of cultural distinctions,
- oversimplification of important content,
- reinforcement of bias due to training data.

There is human review and filtering of rules throughout the project so there is some protection built in, this must continue to be assessed and reviewed as they deploy this across a broader ecosystem of education.

The paradigmatic example of bias in readability systems occurs when models reduce meaningful culturally relevant idiomatic constructs to overly simplified formats. For instance, in Arabic the idiom “ ” (“one hand cannot clap”) carries a culturally implicit meaning about working collaboratively. A naïve simplification model would identify “ ” (“one hand cannot clap”) and reduce it to “you need help.” Thus, the model would lose the richness of the original idiomatic context. To hopefully address the risk of this occurring, we need to include a hybrid review stage that leverages rule-based linguistic resources, which flag idiomatic constructs, and fine-tuned deep learning models that are trained from culturally diverse corpora. This hybrid mechanism will also ensure that readability is not obtained at the loss of meaning or culture, allowing to be ethically accountable for the accessibility issues intended to address in educational NLP, which should be rooted in inclusive and fair practices.

Ethics exist at the forefront of the use of AI-based educational technologies. Not only does algorithmic bias and fairness in profiling learners take centre stage, but including transparency and fairness to facilitate the accessibility and trust of all learners, are just as relevant. Data privacy and the responsible use of language resources are also priority ethical considerations. These ethical aspects contribute towards supporting the long-term sustainable and equitable use of NLP technologies in educational contexts.

Deploying NLP tools in education and learning also entails assessments through good governance and ethical oversight, in addition to technical and pedagogical design as above. Good governance is important to protect students' data from student learning profiles, student performance, privacy legislation, and acts such as General Data Protection Regulation (GDPR). Good governance also encompasses institutional review boards or ethics committees to review the governance of the system design, data use, and assessment decision-making processes. Moreover, other hybrid AI teams can unintentionally replicate inequities by uneven access to infrastructure, training, or devices, giving the impression of supporting equity solutions to sustain the unique needs of an institution, particularly within hybrid or remote higher education. When governance frameworks are cognizant of equity also through the deployment of NLP tools, these frameworks should mean well to support useful approaches to alleviating the digital divide, not sustain it, with useful pedagogy for access to opportunities for more inclusive pathways to learning.

This chapter has principally addressed Arabic, but the approaches we have described clearly have multilingual applications. Lexical simplification, semantic alignment, and data augmentation all have the potential to recognize multilingual, especially low-resource languages. In multilingual contexts, the use of a system can extends the scope of application and open the door for more equitable education across diverse language communities.

9. CONCLUSION AND FUTURE DIRECTIONS

This chapter has presented a thorough and innovative educational project management through artificial intelligence applied to the lexical simplification of Arabic texts. The case study presented showed how the project was able to be technological, pedagogical, and linguistic, and how it could address some of the current challenges of inclusive education. This development is situated within a broader context of digital transformation in education where natural language processing-based tools that are starting to emerge as effective ways to personalize learning, eliminate language barriers and increase educational opportunity.

The project was unique in that it adopted a hybrid approach that combined the strength of (so-called) expert linguistic resources (such as Arabic WordNet, manual lexicons and morphological filters) with the potential of deep learning models (and in particular, both word embeddings and AraBERT). In terms of linguistic ideas, it united two hypothetically mutually-exclusive paradigms (the symbolic approaches of linguistics and the statistical approaches of machine learning). This hybrid approach has a proven track record for meeting a complementary obligation: to simplify Arabic texts for the beginner or upper-beginner or intermediate Arabic learner while maintaining the corresponding meaning, nuances and educational context of the original text.

The empirical findings are clear. Across both automatic metrics (BERTScore and readability scores) and human judgements from experienced teachers, our evaluation of the system established that overall readability for the texts was improved upon. The best-performing hybrid (lexical database + AraBERT) managed to make the texts more simplistic (up to 18.57%) while still exhibiting high similarity in semantics (94.93% F1 score for level 2 texts). This level of performance is impressive, especially in an Arabic language context which is characterized by a high level of morphological richness, diglossia, and a relative lack of annotated resources.

But beyond the statistics, we believe that this project has considerable implications for the educational AI space for a few reasons.

First, it demonstrates that it is possible to design AI systems that are based on the real needs of teachers and learners. One of the key issues with AI projects in education has been the divide between the technical engendering of an educational tool, and the utilization from a pedagogical standpoint. Here, the divide has been possible to bridge as this system was developed utilizing an engaged user-centered design: teachers participated in various aspects of the project - collecting and annotating corpora, building lexicons, and, at the end, evaluating the output of the system.

This active participation ensured the educational relevance of the proposed simplifications and their seamless integration into real educational contexts, whether in the classroom, in self-learning, or in digital environments.

Secondly, the project helps bridge the linguistic digital divide. In Natural Language Processing, the biggest gains in research focus on languages that have sufficient resources, especially English. However, when it comes to Arabic, for example, often treated as an afterthought due to a lack of resources and investment, the challenge is twofold: building successful tools without a lot of data and complicated linguistic specifics. This system fits this logic exactly - it creatively leverages existing resources, enriches that with expert annotation and connects to pre-trained models suitable for the Arabic context, contributing to a degree of normalization in the NLP ecosystem and promoting a type of linguistic fairness in educational technologies.

Thirdly, this project proposes a reproducible model for educational AI project management. From the planning phase to evaluation, including annotation, pipeline simplification development, system architecture, pedagogical validation, and error analysis, each step has been structured according to a clear and agile logic. The use of iterative management through short sprints, cross-validation of data, and rigorous documentation not only allows for the replicability of the project, but also its scalability to other contexts: other language levels, other school subjects, and even other languages. Furthermore, the integration of ethical principles—such as protection against bias, human verification of results, and transparency of methods—reinforces the legitimacy of the approach and its social acceptability in the educational world.

Fourthly, the project offers concrete prospects for large-scale implementation. Several avenues are being considered for the next stages: the development of a personalized simplification system based on learner profiles; the integration of the system into digital learning platforms designed for teaching Arabic as a second language; the extension of simplification from the lexical level to the syntactic and discursive levels; and the construction of a multi-level simplification continuum to support learner progression. All of these developments reinforce the transformative potential of this tool, not only as a technological solution, but also as a lever for educational reform.

In conclusion, this case study exemplifies what careful convergence of language sciences, artificial intelligence, and pedagogy can provide. The project, which is placing users—teachers and learners—at the center of the whole process and drawing from empirical and quantitative studies, all while considering the pedagogical intention of innovation, argues a new way of thinking about education with AI, particularly in low-resource contexts.

In short, this work does not just improve the usability of clarity of Arabic texts for learners. It completely reconfigures the contours of a new model of educational AI projects, based upon inclusion, interdisciplinary collaboration, responsible innovation and their socio-pedagogical impacts. It reminds us that technologies are not the end, but a means to a larger end: to make knowledge more accessible, fairer and more human. From that perspective, Arabic—with all of its beauty and complexity—deserves the investments made in it, and this project is an exemplary demonstration of that.

REFERENCES

A systematic review of literature reviews on artificial intelligence in education (AIED) : A roadmap to a future research agenda | Smart Learning Environments. (s. d.). Consulté 24 août 2025, à l'adresse https://link.springer.com/article/10.1186/s40561-024-00350-5

Al Khalil, M., Habash, N., & Saddiki, H. (2017). Simplification of Arabic Masterpieces for Extensive Reading : A Project Overview. *Procedia Computer Science, 117*, 192–198. DOI: 10.1016/j.procs.2017.10.109

Al-Subaihin, A. A., & Al-Khalifa, H. S. (2011). Al-Baseet : A proposed simplification authoring tool for the Arabic language. *2011 International Conference on Communications and Information Technology (ICCIT)*, 121-125. DOI: 10.1109/ICCITECHNOL.2011.5762661

Al-Thanyyan, S. S., & Azmi, A. M. (2021). Automated Text Simplification : A Survey. ACM Comput. Surv., 54(2), 43:1-43:36. DOI: 10.1145/3442695

Al-Thanyyan, S. S., & Azmi, A. M. (2023). Simplification of Arabic text : A hybrid approach integrating machine translation and transformer-based lexical model. *Journal of King Saud University. Computer and Information Sciences, 35*(8), 101662. DOI: 10.1016/j.jksuci.2023.101662

Alhafni, B., Hazim, R., Liberato, J. P., Khalil, M. A., & Habash, N. (2024). The SAMER Arabic Text Simplification Corpus. arXiv preprint arXiv:2404.18615.

Althobaiti, M., Kruschwitz, U., & Poesio, M. (2014). AraNLP: A Java-based Library for the Processing of Arabic Text. 5.

Antoun, W., Baly, F., & Hajj, H. (2020). Arabert : Transformer-based model for arabic language understanding. arXiv preprint arXiv:2003.00104.

Athiwaratkun, B., Wilson, A. G., & Anandkumar, A. (2018). Probabilistic FastText for Multi-Sense Word Embeddings (No. arXiv:1806.02901). arXiv. /arXiv.1806.02901DOI: 10.18653/v1/P18-1001

Baazeem, I., Al-Khalifa, H., & Al-Salman, A. (2025). AraEyebility : Eye-Tracking Data for Arabic Text Readability. *Computation (Basel, Switzerland), 13*(5), 5. Advance online publication. DOI: 10.3390/computation13050108

Berrichi, S., Nassiri, N., Mazroui, A., & Lakhouaja, A. (2023). Impact of Feature Vectorization Methods on Arabic Text Readability Assessment. In Farhaoui, Y., Rocha, A., Brahmia, Z., & Bhushab, B. (Eds.), *Artificial Intelligence and Smart Environment* (pp. 504–510). Springer International Publishing., DOI: 10.1007/978-3-031-26254-8_73

Biran, O., Brody, S., & Elhadad, N. (2011). Putting it Simply : A Context-Aware Approach to Lexical Simplification. Proceedings of the 49th Annual Meeting of the Association for Computational Linguistics: Human Language Technologies, 496-501. https://aclanthology.org/P11-2087

Black, W., Elkateb, S., Rodriguez, H., Alkhalifa, M., Vossen, P., Pease, A., & Fellbaum, C. (2006). Introducing the Arabic wordnet project. *Proceedings of the third international WordNet conference*, 295-300.

Bott, S., Rello, L., Drndarevic, B., & Saggion, H. (2012). Can Spanish Be Simpler? LexSiS: Lexical Simplification for Spanish. In M. Kay & C. Boitet (Éds.), Proceedings of COLING 2012 (p. 357-374). The COLING 2012 Organizing Committee. https://aclanthology.org/C12-1023

Branch, R. M. (2009). *Instructional Design : The ADDIE Approach.* Springer US., DOI: 10.1007/978-0-387-09506-6

Clark, J. L. D., & Clifford, R. T. (1988). THE FSI/ILR/ACTFL PROFICIENCY SCALES AND TESTING TECHNIQUES : Development, Current Status, and Needed Research. [JSTOR.]. *Studies in Second Language Acquisition, 10*(2), 129–147. DOI: 10.1017/S0272263100007270

De Belder, J., & Moens, M.-F. (2010). Text simplification for children. Prroceedings of the SIGIR workshop on accessible search systems, 19-26.

Elmadani, K. N., Habash, N., & Taha-Thomure, H. (2025). A Large and Balanced Corpus for Fine-grained Arabic Readability Assessment (No. arXiv:2502.13520). arXiv. /arXiv.2502.13520DOI: 10.18653/v1/2025.findings-acl.842

Horn, C., Manduca, C., & Kauchak, D. (2014). Learning a lexical simplifier using wikipedia. *Proceedings of the 52nd Annual Meeting of the Association for Computational Linguistics* (Volume 2: Short Papers), 458-463.

Khallaf, N., & Sharoff, S. (2022, avril 20). Towards Arabic Sentence Simplification via Classification and Generative Approaches. arXiv.Org. https://arxiv.org/abs/2204.09292v1 DOI: 10.18653/v1/2022.wanlp-1.5

Mikolov, T., Chen, K., Corrado, G., & Dean, J. (2013). Efficient estimation of word representations in vector space. arXiv preprint arXiv:1301.3781.

Nassiri, N., Cavalli-Sforza, V., & Lakhouaja, A. (2023). Approaches, Methods, and Resources for Assessing the Readability of Arabic Texts. ACM Transactions on Asian and Low-Resource Language Information Processing, 22(4), 95:1-95:30. DOI: 10.1145/3571510

Nassiri, N., Lakhouaja, A., & Cavalli-Sforza, V. (2018a). Arabic Readability Assessment for Foreign Language Learners. In Silberztein, M., Atigui, F., Kornyshova, E., Métais, E., & Meziane, F. (Eds.), *Natural Language Processing and Information Systems* (pp. 480–488). Springer International Publishing. DOI: 10.1007/978-3-319-91947-8_49

Nassiri, N., Lakhouaja, A., & Cavalli-Sforza, V. (2018b). Modern Standard Arabic Readability Prediction. In Lachkar, A., Bouzoubaa, K., Mazroui, A., Hamdani, A., & Lekhouaja, A. (Eds.), *Arabic Language Processing : From Theory to Practice* (pp. 120–133). Springer International Publishing.

Nassiri, N., Lakhouaja, A., & Cavalli-Sforza, V. (2022a). Arabic L2 readability assessment : Dimensionality reduction study. *Journal of King Saud University. Computer and Information Sciences*, *34*(6, 6, Part B), 3789–3799. DOI: 10.1016/j.jksuci.2020.12.021

Nassiri, N., Lakhouaja, A., & Cavalli-Sforza, V. (2022b). Combining Classical and Non-classical Features to Improve Readability Measures for Arabic First Language Texts. In Kacprzyk, J., Balas, V. E., & Ezziyyani, M. (Eds.), *Advanced Intelligent Systems for Sustainable Development (AI2SD'2020)* (pp. 463–470). Springer International Publishing., DOI: 10.1007/978-3-030-90633-7_39

Paetzold, G. H., & Specia, L. (2017). A Survey on Lexical Simplification. *Journal of Artificial Intelligence Research*, *60*, 549–593. DOI: 10.1613/jair.5526

Pasha, A., Al-Badrashiny, M., Diab, M. T., El Kholy, A., Eskander, R., Habash, N., Pooleery, M., Rambow, O., & Roth, R. (2014). MADAMIRA: A Fast, Comprehensive Tool for Morphological Analysis and Disambiguation of Arabic. *LREC*, *14*, 1094–1101.

Preethi Krishna, P., & Sharada, A. (2020). Word Embeddings—Skip Gram Model. In Gunjan, V. K., Garcia Diaz, V., Cardona, M., Solanki, V. K., & Sunitha, K. V. N. (Eds.), *ICICCT 2019 – System Reliability, Quality Control, Safety, Maintenance and Management* (pp. 133–139). Springer., DOI: 10.1007/978-981-13-8461-5_15

Project Management Institute (Ed.). (2008). *A guide to the project management body of knowledge (PMBOK guide)* (4th ed.). Project Management Institute, Inc.

Qiang, J., Li, Y., Zhu, Y., Yuan, Y., Shi, Y., & Wu, X. (2021). LSBert : Lexical simplification based on BERT. *IEEE/ACM Transactions on Audio, Speech, and Language Processing*, *29*, 3064–3076. DOI: 10.1109/TASLP.2021.3111589

Rhel, H., & Roussinov, D. (2025). Large Language Models and Arabic Content : A Review (No. arXiv:2505.08004). arXiv. https://doi.org//arXiv.2505.08004DOI: 10.48550

Tenney, I., Das, D., & Pavlick, E. (2019). BERT Rediscovers the Classical NLP Pipeline (No. arXiv:1905.05950). arXiv. /arXiv.1905.05950DOI: 10.18653/v1/P19-1452

Wijnands, W., & Stolze, A. (2019). Transforming Education with eduScrum. In Parsons, D., & MacCallum, K. (Eds.), *Agile and Lean Concepts for Teaching and Learning : Bringing Methodologies from Industry to the Classroom* (pp. 95–114). Springer., DOI: 10.1007/978-981-13-2751-3_5

Zhang, T., Kishore, V., Wu, F., Weinberger, K. Q., & Artzi, Y. (2019). Bertscore : Evaluating text generation with bert. arXiv preprint arXiv:1904.09675.

ENDNOTES

1 https://gloss.dliflc.edu/

2 https://learning.aljazeera.net/en

3 https://simple.wikipedia.org/wiki/Main_Page

4 https://huggingface.co/aubmindlab/bert-base-Arabert

Chapter 8
Managing Educational Projects Through the Enunciative Exploration of High School Students' Writings

Salma Karym
https://orcid.org/0009-0009-1107-9053
Faculté des Lettres et Sciences Humaines Fès, Morocco

Samira Hraoui
https://orcid.org/0009-0003-9651-9957
Faculté des Lettres et Sciences Humaines Dhar El Mehraz Fès, Morocco

Rahma Barbara
Faculté des Lettres et Sciences Humaines Dhar El Mehraz Fès, Morocco

Mohamed El Himer
Faculté des Lettres et Sciences Humaines Dhar El Mehraz Fès, Morocco

ABSTRACT

This article covers a problematic concerning most of Moroccan learners of French as a foreign language. It covers one of the four main skills namely writing. Based on a test covering the writing of an intimate letter, this study focuses mainly on the linguistic approach and precisely the enunciative approach. The goal is to see the usage rate of enunciative aspects in students' writings especially since the enunciative system require the presence of several elements. This work aims to show if

DOI: 10.4018/979-8-3373-2975-8.ch008

learners can exploit language course in their writing and if they correctly exploit enunciation components. The sample reveals failures in usage and mastery over enunciation signs in their writing. It appears that the majority are incapable of deploying all of enunciation signs.

INTRODUCTION

In this paper, we focus on one of the four language skills of Moroccan learners: written expression. The focus of this study is the linguistic analysis of French literary works produced by learners of French as a foreign language. Our analysis examines how learners acquire the declarative stance in writing situations.

On the one hand, it is impossible to produce a linguistically correct work without using the declarative stance correctly. In other words, a literary work must contain correct verb conjugation, correctly placed demonstrative markers (e.g., pronouns, time/place references), appropriately chosen modality, and correctly used indirect speech to ensure a high degree of linguistic accuracy.

On the other hand, French is the official first foreign language in Morocco and students start learning French in primary school. In exams, they need to demonstrate a good knowledge of French, especially in written expression. All these reasons explain our choice of this topic.

Our research question is highly relevant and has attracted the interest of many researchers in the field of teaching French as a foreign language (FFL). Our study focuses on several specific research questions related to the use of articulatory devices. The core question can be broken down into several sub-questions.

- In Morocco, the Ministry of Education has taken literacy very seriously and included it in all educational programs. The General Guidelines for Teaching French in High Schools state: “One of the objectives of the reform is to strengthen the language and communication skills of grammar school students. To this end, the curriculum consists of 11 modules covering all disciplines - whether literary, scientific or technical. These modules include:
- *Oral and written language*
- *Literature: the short story and the tale*
- *The literary essay*
- *Argumentation*
- *Poetry*
- *Images and types of discourse*
- *Theater*
- *Full-length literary works*

- *Summary and synthesis of texts*
- *Language for specific purposes*
- *Text analysis*" (M.E.N., 2007, p. 3)

Written performance is therefore always of interest to anyone who wants to assess students' language skills. Even the assessment criteria for French national exams and standardized writing tasks clearly state that half of the grade depends on writing performance, while the other half is distributed among other exam parts.

Given these data, learners of French as a foreign language in Morocco should do their best to perfect their written expression. In order to achieve this, students must first pay more attention to the linguistic aspects of writing. Since most writing tasks focus on narration, description or argumentation, students need to use the phonetic aspects appropriately and recognize the central importance of pronunciation as a linguistic area. Benveniste defines them from the outset as:

"As an individual realization, enunciation can be defined, in relation to language, as a process of appropriation. The speaker appropriates the formal apparatus of the language and expresses their position as a speaker through specific indicators on the one hand, and by means of accessory processes on the other. [...] Every utterance is, explicit or implicit, an allocution; it presupposes an addressee. In enunciation, language is used to express a certain relationship to the world. The very condition of this mobilization of this appropriation of the language lies with the speaker in the need to refer through discourse, and with the other in the possibility of co-referring identically, within the pragmatic consensus that makes each speaker a co-speaker. Reference is an integral part of enunciation" (Benveniste, 1974/1970: 82).

However, even first-year Moroccan high school students do not always produce satisfactory compositions. This raises the question: Do these students' literary works fail because of insufficient use of declarative elements or because of other problems related to writing skills? To answer this question, we conducted a quantitative and qualitative analysis of the use of declarative elements by first-year Moroccan high school students.

Our research questions were divided into three basic research questions regarding the frequency and nature of students' use of declarative elements. These questions can be restated as follows:

- Do first-year baccalaureate students make sufficient use of enunciative forms in their literary compositions?
- Do Moroccan high school students neglect the use of certain enunciative aspects ?
- How do students use these enunciative elements in practice ?

To operationalize our research problem, we will formulate the following specific hypotheses. These reflect our initial concerns regarding the use of enunciative aspects in students' writings:

- Determine whether first-year baccalaureate students resort to enunciative features to enhance the quality of their literary compositions.
- Identify whether some enunciative markers are used more frequently than others.
- Evaluate whether Moroccan high school students use enunciative markers appropriately and effectively.

As was already indicated, if one genuinely wants to create a literary book with a clear structure, the area of enunciation should not be disregarded. Additionally, a lot of researchers have recently been interested in this topic.

We thought it was necessary to carry out a study that combines writing didactics and linguistics. And because enunciation plays a significant part in writing, we decided to choose it as our chosen area of linguistics.

Furthermore, our interest has always been piqued by the ongoing writing challenges encountered by French language learners in Morocco. Therefore, the purpose of this study was to investigate if enunciative traces can be found in students' literary works.

Theoretical features of enunciation and its constituent parts will be presented initially in order to tackle this research challenge. Following a thorough qualitative and interpretive review of the data, we will go into detail about the tools used to acquire the data. Lastly, we will highlight the limits of our study and offer recommendations for repair.

I. ON ENUNCIATION

Reflections on the field of enunciation only became widespread at the end of the 1950s, following the programmatic reflections of Jakobson and Benveniste with his work Problèmes de linguistique générale (Benveniste, 2014). However, this approach had been previously touched upon and observed in several works by Charles Bally and Henry Frei, as revealed by Maingueneau in this quote: “And today we are faced with a vast and active field of research, indeed, but whose various areas are not rigorously articulated with one another,” (Maingueneau, 2023).

Since its inception, the field of enunciation has been understood and described by a number of authors and scholars, including Oswald Ducrot, Antoine Cullioli, Dominique Maingueau, and Emile Benveniste, with whom we observed its emergence. Many people have defined and explained enunciation. The specificity of enunciation, according to Benveniste, is found in the fact that it is "the very act of producing an utterance and not the text of the utterance, which is our object." Ducrot defines it as "the historical event that constitutes, by itself, the appearance of an utterance," (Ducrot, 2014). The speaker is the one who uses words to further his own agenda. While Maingueneau (2023) defines enunciation as "the individual act of using language to oppose it to the utterance, the linguistic object resulting from this use."

Lastly, despite some differences between these definitions, they all agree that enunciation is the process by which a person uses language to create an utterance.

But it appears that every theory takes a different tack when it comes to the study of enunciation. Benveniste (2014) for instance, discusses both discursive enunciation, which includes "all genres in which someone addresses someone else, enunciates as a speaker, and arranges what they say within the category of person," and the formal apparatus of enunciation (Benveniste, 2014).

When "the utterance is entirely devoted to the past, without any intervention of the speaker or any address to the recipient; it is cut off from enunciation; no other indicators of time and place are found except those that situate the event independently of the situation," he also refers to historical enunciation (Benveniste, 2014). However, we come across what Kerbrat-Orecchioni refers to as confined enunciation and extended enunciation.

According to Kerbrat-Orecchioni (2009), the linguistics of enunciation here aims to represent the various relationships between the utterance and the elements constituting the enunciative framework, such as spatio-temporal indices, the enunciator, and the recipient, along with the formalities of the production/reception of the message. This is why the first enunciation is approached extensively, as these terms demonstrate. According to Kerbrat-Orecchioni (2009), the second enunciation is "conceived restrictively, where the linguistics of enunciation only concentrates on one of the essential components of the enunciative framework: the speaker-writer."

However, regardless of the angle from which each researcher views the notion of enunciation and regardless of the differences in their approaches and analyses, no one can deny that "the notion of enunciation is inseparable from that of good formation of an utterance," (Maingueneau, 2023). This leads to the conclusion that enunciation and the utterance are mutually connected, and that every enunciator must rely on enunciative aspects (modalities, indexicals, and reported speech) to anchor their utterance in a given enunciative situation, position themselves relative to its content, ensure the eloquence of their discourse or text, and assume a dual role consisting of enunciating and "taking into account other people, positioning oneself in relation to them, and constructing one's discourse based on this necessary component," (Maingueneau, 2023).

1.1. Utterance / Enunciation

Since enunciation, as previously described, is the individual act of mobilizing language to produce an utterance, there has always been a reciprocal relationship between enunciation and the utterance. On the other hand, it is evident that enunciation produced the speech. Furthermore, only through its output—the utterance—can linguists explain meaning if "the notion of enunciation is appropriate both for characterizing the meaning of utterances (taken as the fact, the given, to be explained) and for establishing the meaning of sentences"[11] (Maingueneau, 2023).

1.2. Modalities

In order to position themselves either in relation to the content of their utterance (statement modalities) or in relation to their co-enunciator (enunciation modalities), the enunciator uses a set of linguistic resources known as modalities to modulate their speech and express their attitude. A detailed examination of a particular discourse type—perhaps scientific or journalistic articles—reveals a unique modal configuration that, in Pollet's words, is as follows: "The article is characterized by more frequent logical modalities, rare subjective statement modalities, and essentially assertive enunciation modalities," (Pollet, 2014). Interestingly, the author's caution is reflected in both the increased number of logical modalities and the rarity of subjective modalities. The author of the citation describes enunciation modalities that may be assertive, subjective, or logical, with the aim of persuading the interlocutor.

This observation first emphasizes the logical predominance of assertive modalities, as the primary goal of this type of text is often to inform or analyze, which pertains to the act of assertion. The quote then contrasts the low presence of subjective statement modalities—those that convey an emotional or appreciative personal judgment, often expressed through adverbs (fortunately, unfortunately) or expressive punctuation—with the higher frequency of logical modalities. The latter, which concern the degree of certainty, probability, or necessity of the content (epistemic or deontic), are often expressed through adverbs (probably, certainly) but also, as the complementary observation suggests, through verbal or adverbial phrases such as "it seems that," "generally speaking," "this tends to indicate that."

The interpretation proposed in the quotation is that this configuration—assertion as a framework, limited direct subjective expression, and frequent use of logical nuance—reflects the author's caution. Rather than asserting categorical truths, the author prefers to mitigate the strength of their claims, thus demonstrating a certain intellectual rigor or a desire for distancing. It appears that subjective modalities are most often represented by adverbs or punctuation. In contrast, logical modalities are more numerous and frequently appear in the form of phrases or expressions such as: "it seems that," "generally speaking," "this tends to indicate that..."

1.3. Deictic Markers

A fundamental linguistic category, deictic markers (also known as "shifters") serve the primary purpose of anchoring the utterance in its enunciation setting.

"Some classes of linguistic features included in the utterance serve to'reflect' its enunciation, integrating aspects of the enunciative context," Maingueneau states, highlighting their integrative importance. These components are essential to the meaning of the statement, and if one wants to comprehend that meaning, one cannot disregard what they allude to.

"The broad meaning of a deictic marker cannot be determined apart from a reference to the message," according to Maingueneau (2023).

While deictic markers include both personal pronouns and spatio-temporal markers, deictic elements do help reveal the meaning of an utterance and highlight clues related to the enunciative context, and these deictic expressions can only be fully explained when we are aware of the particular enunciation situation in which the utterance was produced.

1.3.1. Personal Pronouns

"Indeed, unlike I-you, he is a pronoun in the strict sense, that is to say, an anaphoric element that replaces a noun phrase from which it derives its reference and which was previously introduced into the discourse," stated Maingueneau, confirming that personal markers form the organizing pivot of verbal interaction at the core of the enunciative apparatus (15). Both the speaker, the instance that generates the utterance (I), and the addressee, the instance to which this utterance is directed (you), are deictically referred to by the fundamental pair I-you (or you formal). Unlike the third person (he/she/they), which Benveniste referred to as a "non-person" because it refers to entities located outside the direct sphere of the speech act—objects of discourse rather than active participants—these two markers create the fundamental intersubjective relationship of discourse (Maingueneau, 2023).

However, Dominique Maingueneau emphasizes the adaptability of this system and its modulation possibilities, noting in particular the striking extension of the use of the first and second persons to speechless entities, like infants or domestic animals, in anthropomorphic projection or fictional address. Furthermore, paradigmatic substitutions are common for pragmatic and social reasons: the use of the polite you instead of the informal you, and the traditional replacement of I with the royal we (pluralis majestatis) or the modest we, reflect specific social relationships, establishing distance or respect.

The non-person (he/she), on the other hand, can be used to address a lower-status interlocutor (a sovereign speaking to a subject) or, less frequently but nonetheless, a subordinate speaking to a superior (a servant to their master) in situations of strong hierarchical asymmetry. This depersonalization or objectification represents the highest level of status-based distance. These variations demonstrate how the use of personal identifiers becomes a subtle indicator of interpersonal relationships and socio-discursive practices, going beyond mere reference to the enunciative actants.

1.3.2. Temporal Deictics

Based on Maingueneau's work, temporal markers can be classified according to four temporal relationships with the moment of enunciation, when the point of reference is the current moment:

- Simultaneity: at this moment, at this time, currently, today, now…
- Posteriority: tomorrow, soon, immediately, shortly, in two days, next month, next year…
- Anteriority: yesterday, formerly, once, last Monday, last month, last year…

When the temporal reference is distinct from the moment of enunciation, Maingueneau proposes three other temporal relations:

- Coincidence with the reference point: then, that day, on that Monday...
- Anteriority to the reference point: the day before, a year earlier, a day before, two days earlier...
- Posteriority to the reference point: the next day, two days later, a day after, the following month...

However, Maingueneau contends that in this second instance, these parts are non-deictic and not deictics because they belong to a different temporal reference point rather than the speaker's present.

1.3.3. Spatial Deictics

The category of spatial deictics, according to Maingueneau (2023), is a diverse and intricate collection of linguistic instruments whose main purpose is to ground the utterance in a particular spatial arrangement that is inextricably linked to the enunciation context. These indicators include a variety of grammatical resources that collectively provide the spatial context of speech, rather than just a list of locative adverbs. Maingueneau (2023) highlights the crucial function of demonstratives, whether they are pronouns (this one, that one, this, that) or determiners (this, that, these). These components facilitate ostensive designation, which establishes a direct link between language and the observed or evoked world by pointing to objects that are either prominently featured by the discursive framework or present in the interlocutors' immediate surroundings.

Presentative forms like "here is" and "there is" (voici / voilà) also have a distinct and dynamic function. They help to grab the addressee's attention by introducing new referents in a noticeable manner, which goes beyond simple placement. They function as a form of language staging, indicating the emergence or emphasis of a component that the speaker wishes to draw attention to. This capacity to direct the addressee's cognitive focus is essential for managing shared information and advancing the conversation.

Finally, and more prototypically, Maingueneau (2023) includes adverbial elements and adverbial phrases that structure space around the speaking subject, considered the origin of spatial referencing—the zero point or origo of spatial deixis. These adverbs (here, there, over there) and phrases (near, far from, in front of, behind, to the left, to the right, above, below) do not simply indicate a place in absolute terms. They position entities and events in relation to each other and, fundamentally, in relation to the speaker's location at the moment of speaking. They thus map out a network of spatial relationships based on key axes: proximity and distance in relation to the ego (here vs. there/over there), lateral and front-back orientation (to the left/ right, in front/behind), and verticality (above/below).

According to Maingueneau (2023), spatial deictics is an essential system that enables language to interact with the environment around the act of communication. They are not only locative labels; rather, they are necessary operators that actively create the enunciative scene, place discourse participants and objects in a common frame of reference, and represent the speaking subject's own perspective on how space is perceived and organized. For referential interpretation and the pragmatic consistency of any verbal encounter, their utilization is consequently essential.

1.4. Verb Tenses

The temporal structuring of events constitutes a fundamental pillar of discourse organization. The positioning of reported facts along the timeline—whether in the present, past, or future—explains the near-systematic predominance of the indicative mood, the mode of assertion and temporal localization par excellence. According to the foundational distinction established by Émile Benveniste between the two planes of enunciation, this temporal management manifests differently. Within the framework of discursive enunciation (or "discourse"), which anchors the utterance in the communication situation and the subjectivity of the speaker (the "I-here-now"), the present coincides with the very moment of enunciation, the simple future projects the event beyond that moment, and the present perfect refers to past events still relevant to the speaker's present. These temporal choices characterize direct exchanges, letters, and memoirs, where the enunciative instance is strongly present.

Historical enunciation (also known as "narrative"), in stark contrast to this method, creates a temporal framework that is independent of the enunciation process. Here, historical events are described as though they were happening objectively, without specifically mentioning the narrator's current point of view. This plane promotes certain tenses: the pluperfect to indicate anteriority in relation to another instant in the past, the imperfect for background descriptions or continuous actions, and the simple past for major events perceived as finished and punctual. The conditional (present or past) is used to convey a future concept that is imagined from a point in the past; this is known as the "future in the past." According to Benveniste, this temporal opposition is a key indicator that significantly separates these two basic approaches to language use and world narration.

1.5 Reported Speech

Maingueneau asserts that humans have an innate tendency to assume that an utterance is a uniform sequence generated by a single enunciator in a particular enunciation scenario. The enunciator, however, is forced to record statements made by themselves or by another speaker in a different enunciative circumstance because "in actuality, speech is continually traversed by what has already been said, and at times by what is still to be uttered."[17]. To put it another way, all discourses have references to and echoes of earlier discourses. It is important to recognize that social and cultural elements influence the discursive relationship that exists.

As Maingueneau (2023) highlights in the following sentence, "Quoting one utterance within another is not a simple phenomena," the presence of other voices within a discourse is especially significant and difficult. It has long been understood in French that there are three different "strategies" for quotation, each with unique characteristics based on the kind of interaction formed between the quoting discourse (QDis) and the quoted discourse (QDed).

- There are three forms of reported speech:
- Direct speech (DS) preserves the independence of the QDed with respect to the QDis.
- Indirect speech (IS) removes all autonomy from the QDed and subordinates it to the act of enunciation of the QDis.

In a unique enunciative mode, free indirect speech (FIS), which is mostly employed in written language, especially in literary narration, blends the characteristics of both DS and IS to varied degrees.

We can confirm that enunciation emerges as a process inseparable from the utterance based on our discussion of enunciation and its components in the first section and the presentation that follows. It helps the enunciator place oneself in relation to the substance of the utterance, anchor it in a particular enunciative circumstance, and assume a dual function.

In the study of enunciative systems, the different enunciative features that were previously discussed make up essential mechanisms. It's also important to note that these factors bring up interesting and recurrent questions for this field of study.

2. METHODOLOGICAL FRAMEWORK AND DATA COLLECTION

2.1. Research Variables

Our method is predicated on the examination of two variables: the written works of the pupils and their enunciation. The enunciative features include verb tenses, reported speech, spatial and temporal deictics, personal pronouns, enunciation modalities, and twenty written products. All of these factors will aid in our response to the research questions and enable us to confirm whether or not Moroccan high school students' literary compositions have enunciative elements and whether or not these enunciative features are employed correctly when creating literary works.

The main goal of this study is to provide us a well-informed viewpoint on a key question: do Moroccan high school students actually apply all of the language and literary skills they have been taught during their education? More precisely, we want to know if the information they learn in language classes, the advantages of reading for pleasure outside of school, and the introduction to literary text analysis they receive as early as middle school, all significantly affect the caliber of their own literary writing.

Our methodological decision concentrated on the particular examination of enunciative components in order to comprehend this intricate dynamic. This aspect of language is very instructive since it is not learned in a single block; rather, it shows how well a student can mobilize and integrate different concepts and processes that have been studied gradually and independently in a logical way. The learner's stance and discourse construction are reflected in their enunciation.

Therefore, it would tend to indicate a good assimilation and successful application of their information if our observations show that the students are using enunciative markers in a meaningful and effective way. On the other hand, obvious problems with the prudent application of these elements may indicate that there are still issues with the conceptualization of these ideas or that there are gaps in the actual learning process. Therefore, in addition to its main goal, this study may open the door for further extensive investigations. It might inspire us to investigate other topics, namely if the language skills that are formally taught in secondary school are actually learned and, more importantly, whether Moroccan high school students are using them appropriately in real-world writing assignments.

2.2. Sample

We concentrated on a fairly narrow set of Moroccan high school students who were enrolled in the First Year of the Baccalaureate for the 2023–2024 academic year in order to carry out our research successfully. This is not a simple decision. We begin with the logical premise that these young people should have a strong writing ability by the time they reach this point in their journey, especially when it comes to literary literature. Our worries and research inquiries are focused on this ability specifically. As a result, we anticipate that these students will not only possess a strong command of the French language and communication skills, but also be somewhat comfortable with a variety of writing instruments and procedures.

There are two fundamental foundations around which this expectation—or, maybe more accurately, this working hypothesis—is built. On the one hand, their previous education, which they have accrued throughout middle school and their sophomore year, is especially tailored to provide them all the information and abilities needed to write in French in an organized, cohesive, and complex manner. Typically, they have already discussed writing skills, grammar, vocabulary, and text analysis. The standardized regional exam that they must take at the end of the year, however, is a very tangible component that supports this notion. It's interesting to notice that written production plays a significant role in this exam, which is a crucial point in their education.

But we have to be practical. There are a significant number of pupils participating nationwide. Setting up a study that would involve this entire large population well surpassed our capabilities, both logistically and monetarily, as well as in terms of time. Performing a thorough survey was just not possible. Given this inevitable material limitation, it became clear that we needed to approach this group in a different way by focusing on a representative sample. For this reason, we decided to create a sample. We chose twenty pupils at random to accomplish this. All of these participants are from Lycée 2 Octobre, which is situated close to the Boulmane provincial delegation.

2.3. Research Instruments

To conduct this study, we chose a writing assignment as our research tool, which would give us literary works authored by the aforementioned sample. The personal letter is the text type that was selected. The fact that the letter genre has several discursive and enunciative components that enable us to carry out this study effectively supports our selection. A personal letter was required of every student in our sample. Thus, "create a text in which you share with your audience personal facts, an experience, and feelings" is one of the components of the subject statement. Make an effort to use the various enunciative elements in your writing. However, this genre represents an easy-to-complete and easy-to-analyse work.

Additionally, a personal letter gives the writer the opportunity to recount, explain, report, debate, seek for information, or even give credit to someone or something. As a result, the subject and literary writing style are up to the student. As scholars examining these texts, we are more likely to come across a diverse range of enunciative elements. This practice will also give us a manageable and interesting corpus, which will help us accomplish our goals without getting bored and making the writing and analysis less complicated.

We created an assessment grid to go along with each production in order to recognize and analyze the enunciative elements included in these texts and determine whether they are used correctly.

With the use of this grid, we can: Determine which enunciative elements pupils use most frequently and which do not. Find out how each writing makes use of these elements. Our study will do this by using both a qualitative and quantitative component.

We shall first ascertain whether each enunciative component is present or absent for each student's written work. We shall note that the aspect is not used if it is not present. We will evaluate its usage if it exists. We shall indicate "misuse" when the aspect is used wrongly.

"Fair use" will be noted if the aspect is typically used incorrectly but occasionally correctly. We shall record "good use" if it is applied correctly overall. Lastly, we will mark "very good use" if the enunciative element is employed correctly and consistently throughout the text.

Table 1. student copy assessment tool

Appreciations Aspects	No use	Bad use	Fair use	Good use	Very good use
Personal Pronouns					
Deictic Markers					
Temporal Deictics					
Verb Tenses					
Modalities					
Reported Speech					

2.4. Data Collection

We were able to proceed to the vital phase of gathering field data in April. We were really lucky to receive a warm welcome at Lycée 2 Octobre in Imouzzer Marmoucha, which made this possible. We had a lot of help from the principal of the school and an on-site French teacher. Their invaluable cooperation allowed us to set up a meeting with a group of first-year high school students—exactly the demographic we were looking for. We were able to introduce the task we had planned for our study to them.

In addition, one of the high school's general supervisors had a significant role in the practical organization. As they say, he "stretched out" to us by assigning us a particular room. This made it possible for us to administer the test in a setting that was optimal for the students' concentration: quiet and free from the typical commotion.

We were adamant about avoiding interfering with the teacher's and his students' already hectic schedules or the regular operation of the classrooms. We decided to schedule this session during a free period, when the students were available, for that reason. The chief supervisor's engagement was outstanding; he made sure that everything was done with the highest seriousness. We also need to emphasize the pupils' attitude, which was one of understanding and cooperation. As a result, the test proceeded without incident, following the guidelines, and nobody tried to replicate or assist one another.

2.5. Data Analysis

Excel and SPSS software were used to handle the data that was gathered, even though the assessment grid assisted us in identifying the enunciative elements that were employed and categorizing them based on the quality of their usage. These computations' outcomes enabled us to create graphs and significantly simplified the analysis.

In the parts that follow, we will go into more detail and explain the data that was gathered and how it was analyzed.

2.6. Interpretation and Analysis of the Data

We share the data that has been gathered at this point in our investigation. We used graphs and tables to accomplish this. The percentage and frequency of the enunciative elements found in the students' papers, along with the caliber of their use, are displayed in this presentation. There is a related graph for every table. Stated differently, we give a table that illustrates the effectiveness of each enunciative element, followed by a graph that graphically depicts the data. Additionally, there is a single comprehensive table and accompanying graph that show how all enunciative elements are used overall in the students' written work.

We show seven tables and seven graphs in all. This structure was chosen to improve the analysis's clarity, make comparisons easier, and make it easier to interpret the results.

As mentioned before, first-year baccalaureate students are the study's participants. Since there are twenty pupils in all, we are examining twenty distinct literary works, each of which has its unique enunciative markers.

2.6.1. Rate of Use of Enunciative Aspects

Table 2.

Aspects enunciative	The tenses verbs	The people	Deictic Markers	Temporal Deictics	The modalities	The Reported Speech
Number of copies comprising each Aspect	20	20	9	18	11	0
The percentages	100%	100%	45%	90%	55%	0%

Table 2 lists the quantity and proportion of written works that use each enunciative feature. The number of pupils who employed each enunciative aspect is shown in Table 2.

Additionally, it displays the enunciative features that firstyear baccalaureate students primarily use, infrequently use, and never use. Stated differently, this table enables us to pinpoint the enunciative elements that first-year baccalaureate students use to improve their literary writing.

To better decode and explain the results, the data in the above table is converted into graphs in the following graphic.

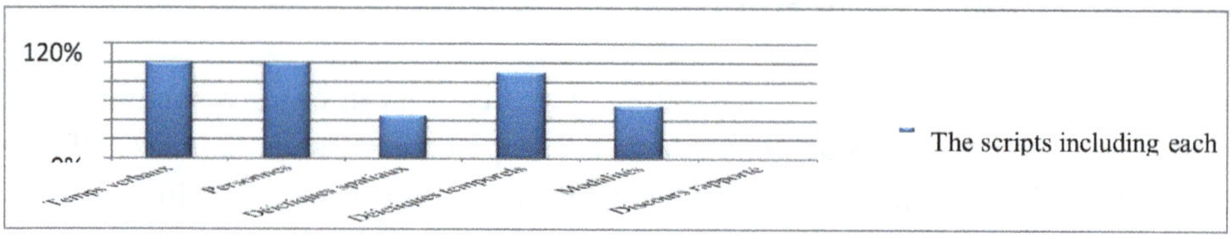

The data from Table 1 is represented by the chart in Diagram 1. It displays the proportion of papers that contain each enunciative feature. This figure makes it clear which enunciative elements first-year baccalaureate students use most frequently and which ones they use less. In contrast to reported speech, which is absent from all of the papers, we can thus see right away that all of the papers use personal pronouns and verb tenses. This suggests that while these kids totally disregard the usage of reported speech, they utilize personal pronouns and are unable to write without using verb tenses.

2.6.2. Quality of Use of Each Enunciative Aspect

The purpose of this study is to ascertain if first-year baccalaureate students use enunciative elements to improve their written literary creations, as well as whether they do so in a way that truly improves their writing. This prompts us to assess how well the students' papers employ enunciative elements. In order to conduct this study, we looked at each enunciative component separately.

2.6.3. Verb Tenses

Table 3. The use of verb tenses in students' written productions

Appreciations	No use	Bad use	Fair use	Good use	Very Good use
Number of copies	0	6	6	6	2
The percentages	0%	30%	30%	30%	10%

Writing a literary work is undoubtedly impossible without employing verb tenses and conjugating verbs. Verb tenses aren't usually used correctly, though. Few students adhere to verb tense agreement and employ tenses correctly, as Table 3 demonstrates. Just 10% of our sample chose the appropriate verb tenses, as shown in this very table.

The many verb tenses used by first-year baccalaureate students in Morocco are clearly distinguished in the accompanying figure.

Figure 2. The use of verb tenses

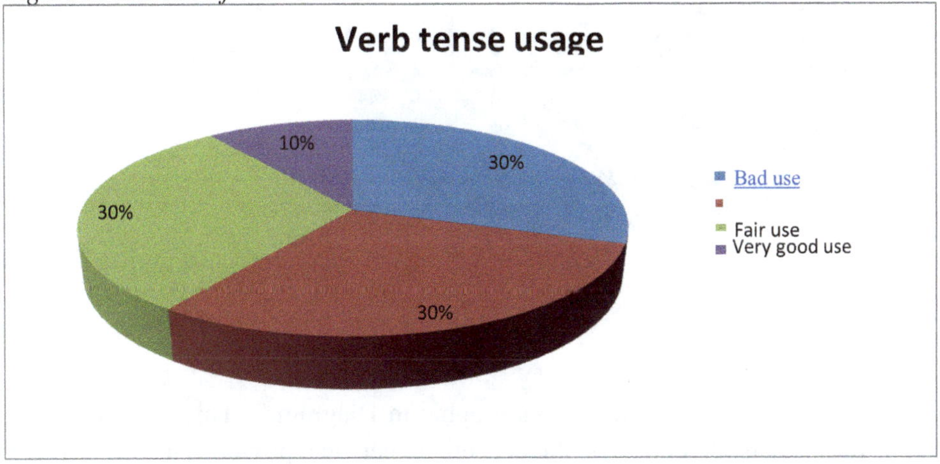

Figure 2 shows that Moroccan high school students use verb tenses differently, with just 10% of our sample exhibiting very good verb tenses usage.

2.6.4. Personal Pronouns

Table 4. The use of personal pronouns in students' papers

Appreciations	No use	Bad use	Fair Use	Good use	Very Good use
Number of copies	0	1	2	2	15
The percentages	0%	5%	10%	10%	75%

Table 4 displays the quantity and proportion of personal pronouns used by first-year baccalaureate students based on how well they utilize them. We may deduce right away that all students, without exception, employ personal pronouns in their written literary products because no absence of use is found in any of the papers.

Figure 3. The use of personal pronouns in students' papers

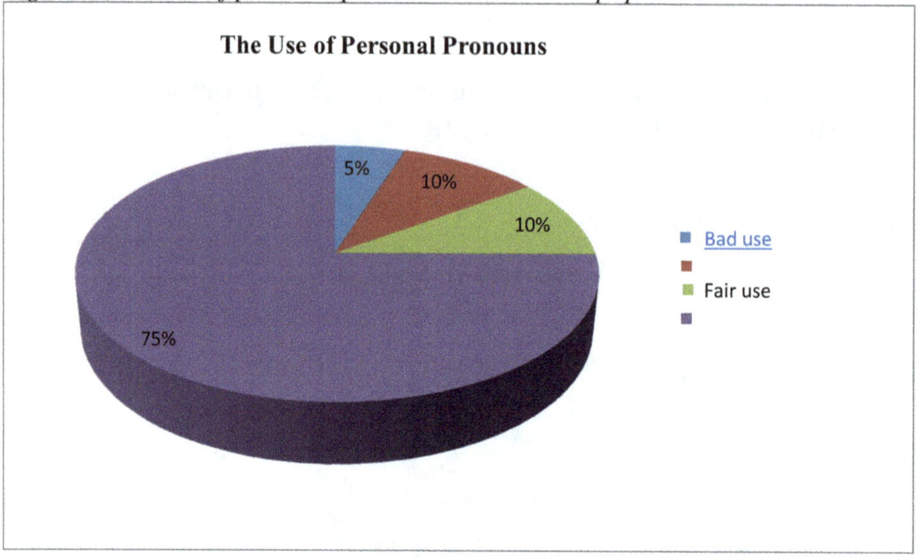

Table 5's data is converted into a graphic in Diagram 3. This graphic shows that while a small percentage of students utilize personal pronouns incorrectly, the majority of students use them quite well in their literary compositions.

2.6.5 .Spatial Deictics

Table 5. The use of spatial deictics in students' papers

Appreciations	No use	Bad use	Fair use	Good use	Very good use
Number of copies	11	0	5	3	1
The percentages	55%	0%	25%	15%	5%

The quality of spatial deictics usage among Moroccan first-year baccalaureate students is seen in Table 5. Table 5 reveals that around half of the students do not employ spatial deictics, despite the fact that they are crucial in literary writing for indicating spatial placement or highlighting a fact or item.

To provide more clarity and aid in the interpretation of the data, the diagram below transforms the information from Table 4 into a graphical manner.

By showing that 55% of students did not apply spatial deictics, Diagram 4, which displays the data from Table 4, supports our previous findings. Only 5% of the minority of students who did use spatial deictics made excellent use of these enunciative features, according to the same graphic.

Figure 4. The use of spatial deictics in students' papers

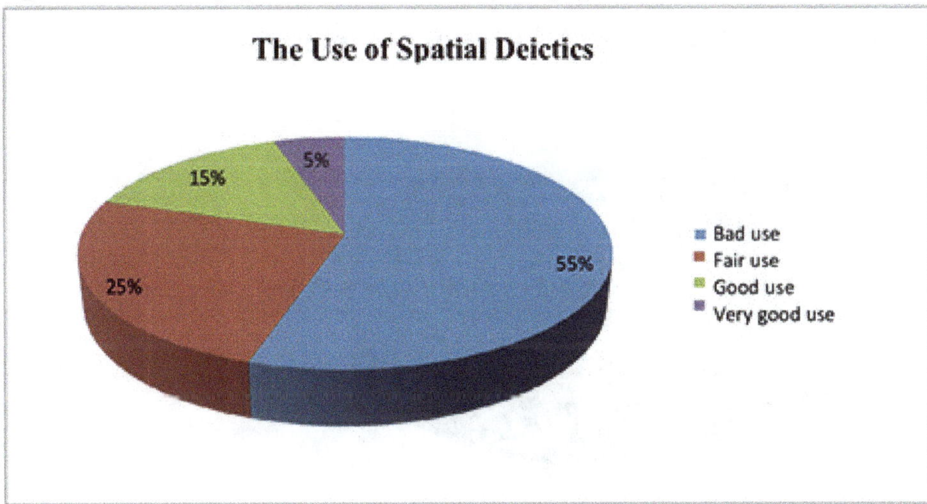

Figure 4, which presents the data from Table 5, confirms our earlier observations by indicating that 55% of students did not use spatial deictics. This same diagram also reveals that among the minority of students who did employ spatial deictics, only 5% used these enunciative aspects very effectively.

2.6.6. Temporal Deictics

Table 6. The use of temporal deictics in students' papers

Appreciations	No use	Bad use	Fair use	Good use	Very good use
Number of copies	2	2	11	4	1
The percentages	10%	10%	55%	20%	5%

The data illustrating the usage of temporal deictics seen in the student articles we assessed is shown in Table 6. In literary writing, temporal deictics is just as important as spatial deictics. The fact that 90% of pupils use this enunciative feature shows that they are aware of this and frequently use it.

However, just 5% of the students who employ this enunciative element in their papers showed very good usage, while 55% of them only used it reasonably well.

The data from Table 6 is displayed graphically in the following picture, which makes the figures in Table 6 more understandable.

Figure 5. The use of temporal deictics in students' papers

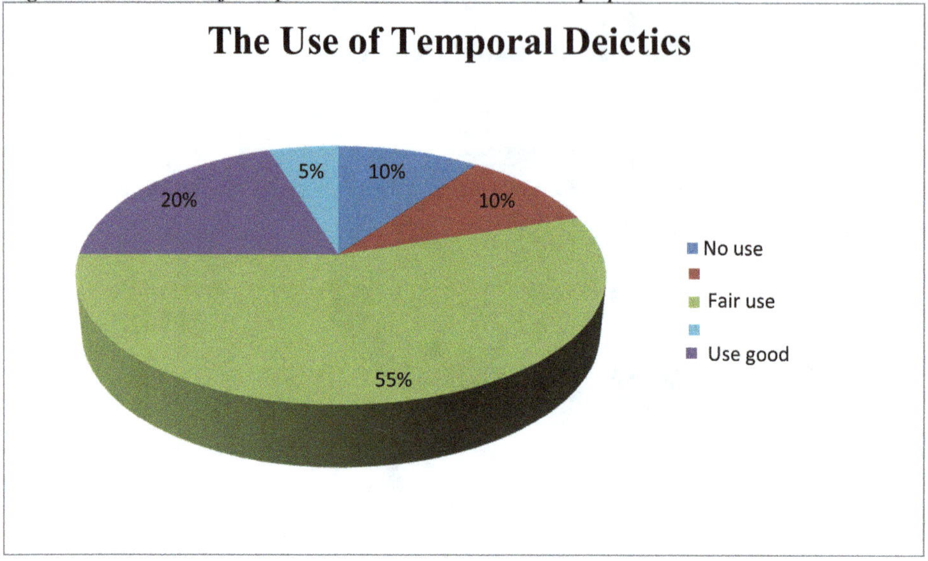

2.6.7. Modalities

Table 7. The use of modalities in students' papers

Appreciations	No use	Bad use	Fair use	Good use	Very good use
Number of copies	9	2	9	0	0
The percentages	45%	10%	45%	0%	0%

The statistics showing how modalities were used in first-year baccalaureate students' papers is shown in Table 7. Modalities, whether factual or subjective, are used to provide subtlety or distancing from the speech. We discovered that almost half of the students who were assessed did not make use of these modalities. The other half of our sample knows the importance of modalities and makes an effort to employ them. None of the students, however, used this enunciative feature well or very well, as shown by the data in Table 7.

Figure 6. The use of modalities in students' papers

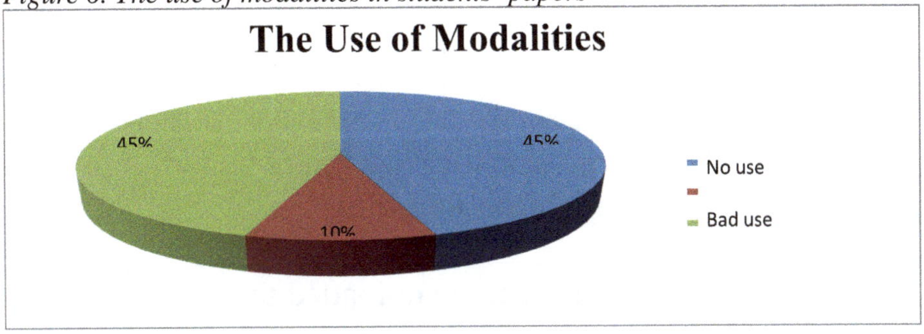

Figure 6 presents the data from Table 8, illustrating the use of modalities observed in the literary compositions of Moroccan high school students enrolled in the first year of the baccalaureate program.

2.6.8. Reported Speech

Table 8. The use of reported speech in students' papers

Appreciations	No use	Bad use	Fair use	Good use	Very good use
Number of copies	20	0	0	0	0
The percentages	100%	0%	0%	0%	0%

Figure 7. The use of reported speech in students' papers

Please provide missing figure by
proofing deadline.

New files can be uploaded to
the proofing system via
comment attachment or emailed to
bookproofing@igi-global.com

None of the first-year baccalaureate students included reported speech in their creative productions, as seen in Table 7 and Diagram 7. This omission implies a lack of knowledge or assurance on this enunciative feature.

We have provided statistical information gathered from student papers throughout this investigation. We sought to compare usage rates across different enunciative elements and investigate how Moroccan first-year baccalaureate students used them by employing numerical values and percentages. To create lucid and well-structured visuals, the material was arranged into tables and diagrams.

3. COMPARATIVE ANALYSIS OF RESULTS

In this section, we will go into more detail about the preliminary remarks we made after presenting the quantitative and qualitative data on the use of enunciative elements and how they were deployed in the papers of first-year baccalaureate students in Morocco. With the goal of addressing our research questions, this section will be devoted to the data analysis and comparison.

We will take the following steps in order to compare the outcomes:

We shall start by comparing how frequently enunciative elements are used. The purpose of this comparison is to ascertain whether Moroccan high school pupils employ all enunciative elements or if some are neglected.

We will then examine each enunciative aspect's usage separately. At this point, we will evaluate if Moroccan first-year baccalaureate students employ these elements appropriately or if they include them into their literary works without completely grasping how to use them or their purposes.

3.1. Comparison of Usage Rates of Enunciative Aspects

Table 2 and Figure 1 show that different enunciative characteristics are used at different rates. Not all enunciative elements are used in every literary work by Moroccan high school pupils, as the figures, percentages, and columns of the histogram demonstrate.

Personal pronouns and verb tenses rank highest on the list, appearing in every paper. Temporal deictics, which appear in 90% of the writings, come in second. After spatial deictics, which were found in just 45% of the compositions, the use of modalities ranks third with 55% of the papers. Lastly, none of the literary works make use of reported speech.

3.2. Comparison of How Each Enunciative Aspect Is Used

3.2.1. Use of Verb Tenses

The majority of Moroccan pupils have trouble with tense agreement and word conjugation. Students frequently neglect to use the proper tense for their statements, despite the fact that verb tenses are employed consistently in their writing. In particular, thirty percent of the papers show inadequate command of verb tenses, thirty percent show a decent degree of competency, and thirty percent show an excellent level. Verb tenses are used very well in just 10% of the students' writings. Table 3 and the accompanying figure support these conclusions.

3.2.2. Use of Personal Pronouns

We noticed that every student's work used personal pronouns. The majority of first-year baccalaureate students used personal pronouns quite well, as seen in Table 4 and Figure 3. In the meantime, 10% of the papers had both good and fairly good utilization. Last but not least, just 5% of the papers had improper personal pronoun usage.

3.2.3. Use of Spatial Deictics

The restricted usage of spatial deictics by a minority of our sample (less than 50%) is confirmed by Table 4 and Figure 4. The distribution of usage quality among individuals who used this enunciative aspect is as follows: Just 5% showed extremely good mastery, 15% used it well, and 25% used it very well.

3.2.4. Use of Temporal Deictics

Only ten percent of pupils did not apply temporal deictics, as shown in Table 6 and Figure 5. 20% of students utilized them well, 55% used them fairly well, and 10% used them incorrectly. Only the remaining 5% of pupils demonstrated excellent usage of temporal deictics.

3.2.5. Use of Modalities

According to Table 7 and Figure 6, 45% of first-year baccalaureate students do not employ modalities in their creative compositions, 10% do so in a poor manner, and the remaining 45% are unable to employ modalities above a very acceptable level.

3.2.6. Use of Reported Speech

Figure 7 demonstrates that none of the students' literary works used reported speech. Stated differently, none of the baccalaureate first-year students used reported speech in their work. For this reason, a comparative analysis is not possible.

This absence implies that, although though reported speech is taught at several educational levels, such as primary, middle, and high school, it presents a substantial obstacle for students. The necessary modifications required for reported speech, such as modifying verb tenses, spatiotemporal markers, and deictic phrases, can present challenges for learners.

According to the comparison, first-year Moroccan high school students do not employ all enunciative features, and their usage rate does not always indicate whether they are used well or poorly. Put another way, a lot of students' work does not always contain the most commonly used feature, and a number of students can use an enunciative aspect quite successfully without it showing up in every paper.

Additionally, it appears that many students utilize enunciative elements in their writing without truly understanding how to use them. We can therefore draw the conclusion that only a small percentage of students frequently make excellent use of enunciative elements. The majority of students simply employ personal pronouns correctly and consistently, making them the only linguistic ingredient.

Additionally, statistical data analysis and comparison show that no student uses all enunciative characteristics well or very well. Additionally, none of the first-year baccalaureate students have used reported speech in their literary productions, thus they all struggle with it.

It's crucial to keep in mind that our goal was to address the hypotheses that were developed at the start of this study, even after the analysis, comparisons, and data interpretation are complete.

The findings demonstrated that each enunciative feature was used at different frequencies and in various ways in the written works of first-year baccalaureate students.

The limitations of our study are discussed in this final section, along with some recommendations to assist address the enunciative deficiencies seen in students' essays.

4. SUGGESTIONS, REMEDIAL MEASURES, AND LIMITATIONS OF OUR STUDY

We offer a number of recommendations to solve the enunciation issues among Moroccan high school students enrolled in the first year of the baccalaureate, based on the enunciative technique we took to their literary works.

Our suggested fixes address the foreign language learners, teachers, and the taught content.

Moroccan high school students are encouraged to pay great attention to the literary texts they meet in order to overcome enunciation issues and master the usage of enunciative elements, which will make them tools for developing their written literary outputs.

When reading, these students need to pay closer attention to the arrangement of enunciative elements, how they are used, and most importantly, the agreement of verb tenses.

Language instruction should be taken seriously by first-year Moroccan baccalaureate students. To put it another way, they need to understand that language instruction is always meant to help students write better.

In turn, teachers of French as a foreign language are urged to employ writing exercises to remind students of the vital function language plays and the significance of its different parts in effective literary writing. On occasion, these teachers might even give students handouts to go over and reinforce language components, especially enunciative markers.

Moroccan baccalaureate students in their first year should take language teaching seriously. In other words, they must realize that the goal of language training is always to improve students' writing.

Teachers of French as a foreign language are therefore encouraged to use writing activities to remind students of the importance of language and the role that its many components play in producing literary writing that is effective. These teachers may occasionally even provide their pupils with handouts to review and reinforce language elements, particularly enunciative markers.

Teachers of French as a foreign language are urged to halt on passages or even sentences during reading exercises to highlight enunciative features and the way authors express and use them. Learners of French as a second language will absorb enunciative elements more readily if they are given in an interesting and appealing way. Additionally, Moroccan pupils will find it easier to remember the applications and locations of enunciative elements if they are given examples that are simpler and less complicated.

It would be better for Moroccan textbooks to go into greater detail on reported speech in order to acquaint students with its use.

4.1 Study Limitations

As with any research project, this study has inherent limitations that need to be clearly stated.

First, we found that there aren't many research explicitly devoted to the formal assessment of enunciative elements in students' written work in the scientific literature now in publication. This made it more difficult for us to contrast our results with a larger and more well-established corpus of work on this particular topic.

Second, from a methodological standpoint, the study's time frame—one semester—did not provide a thorough investigation of the subject.

Third, it must be recognized that the qualitative study of enunciative markers inherently involves some degree of interpretation. Even if we have made an effort to use consistent criteria, despite the methodological rigor we have sought, the identification and classification of some modalities or indications of subjectivity may involve an interpretive dimension that cannot be fully objectified.

Fourth, we only examined enunciative features in our analysis. Though not explicitly examined in this study, the challenges seen in students' writing are frequently multifaceted and may stem from a complex interaction with other linguistic skills (such as syntactic, lexical, or discursive competence).

Lastly, the nature of the gathered corpus may not have consistently used the entire spectrum of possible enunciative tactics, even though it is indicative of common writing activities. For the purpose of more precisely evaluating the acquisition of specific enunciative signals, certain tasks may be required. As a result, these elements restrict the range of our findings and create opportunities for more, more thorough, and extended study.

5. REMEDIATION APPROACHES

Targeted pedagogical reflection is necessary in light of the major challenges that have been identified in relation to reported speech, modalities, and the use of spatial deictics. Even though reported speech was introduced early in the curriculum, many students still find it to be a "stumbling hurdle." This calls into question the suitability or efficacy of the teaching and consolidating approaches used. It is worth taking into account and paying attention to the theory that the method might be overly theoretical or not properly connected to real-world writing assignments.

In order to overcome barriers and automate a more confident and pertinent use of these crucial enunciative tools, we suggest sessions in which students revisit these ideas through a range of activities (contrastive analysis, transformation exercises based on communicative situations, integration into micro-writing tasks). Because of this, our proposal is to implement a three-hour remedial period.

5.1 Remediation 1

- **Activity**: Reading
- **Approach**: Comparative Analysis
- **Material**: Chapter 6, *The Last Day of a Condemned Man*
- **Objective**: To grasp the importance of modality in constructing meaning

Procedure: Remediation Activity: Exploring Modality through Punctuation and Modal Expressions

The convicted man challenges the motivations, style, and method of writing in this book, which is presented as a deliberative monologue. This passage demonstrates a number of modalities, including logical modalities that convey position or detachment, exclamatory and rhetorical questions that reflect shock and rejection of the statement, and questions that indicate hesitation.

We will first eliminate all punctuation and modal statements from the text in order to emphasize the significance of these modalities. After reading the revised version, students will evaluate how the meaning is altered by the removal of these components. Students will next read the original passage again after we have added the missing punctuation and modal phrases. . Finally, they will compare the construction of meaning in both versions to understand the role of modality in conveying the author's intent and emotional nuance.

5.2 Remediation 2

- **Activity**: Speaking
- **Approach**: Role-play
- **Objective**: To master the use of reported speech

This lesson aims to improve students' comprehension and application of reported speech. One kid is momentarily removed from the group at the start of the activity. Two more students converse in the meantime. When the first student returns, a fourth student is assigned to explain to them what was discussed in the prior conversation. This calls for modifying the verb tenses employed as well as the proper sentence kinds, such as imperative, declarative, or interrogative.

This practice aims to actively instill the following qualities in students:

- Recognize that the verb tense is the primary determinant of reporting speech.
- Be aware that the verb tense of the original utterance must be changed in order to accommodate the introductory verb.
- Identify the presence and function of modal expressions.LinkedIn

This role-play will be complemented by a transformation exercise during the appropriation phase to reinforce these concepts.

5.3 Remediation 3

- **Activity**: Guided Practice
- **Material**: Topographic Map
- **Objective**: To read a map and orient oneself within topographic space

Procedure

Students will practice locating themselves and describing positions by using and modifying relevant spatial deictic expressions within their statements while seeing the topographic map on a video projector.

Results

The objective of this focused remediation approach is not just to fix certain mistakes. The main goal is to increase students' awareness of enunciative mechanisms and how they affect communication in order to cultivate in them a long-lasting self-correction dynamic.

The goal of the comparison approach to modalities (Remediation 1) is to deeply comprehend their pragmatic and semantic worth. The goal of the reported speech role-play (Remediation 2) is to turn frequently passive knowledge into an active skill that is aware of enunciation settings. Finally, the use of spatial deictics is grounded in a tangible practice of localization and description in the work that uses a topographic map (Remediation 3).

Combining these methods is intended to provide students a greater understanding of metalanguage so they may apply these abilities to their own written work. This is especially important in post-baccalaureate administrative correspondence, where referential accuracy, clarity, and pragmatic appropriateness are essential and any ambiguity might have dire repercussions, potentially turning into a "fatal error." This emphasizes how crucial reinforced mastery of enunciative competence is and how useful it is.

CONCLUSION

The results of this study are summarized in this chapter, which also lays out potential directions for further research. Prior to exploring these possibilities, it is crucial to keep in mind the goals that directed our research:

1. To ascertain whether enunciative elements are used by first-year baccalaureate students to improve their written literary productions;
2. To discover whether some enunciative markers are used more frequently than others; and
3. To ascertain whether Moroccan high school students make appropriate use of enunciative markers.

Our study's findings demonstrate that enunciative elements are indeed used in the writing of Moroccan high school freshmen enrolled in their first year of baccalaureate. These kids, however, tend to rely more on particular enunciative markers than others and do not employ all of them. Even the enunciative elements found in their writings are not always employed well. All students employ verb tenses, but only a small percentage use them correctly. Not all student papers contain spatial and temporal deictics, and when they do, they are frequently misused or positioned incorrectly. However, deictics are essential for effective description, geographical localization, and storytelling. . However, it may be said that most Moroccan high school students appropriately use personal pronouns. About half of them use modalities quite well, while the other half either don't use them at all or misuse them. Lastly, it's crucial to remember that none of the kids' written works included reported speech.

First, the study's findings make us wonder if the French language classes offered to Moroccan students actually help them improve their language skills. This forces us to investigate a different problem and a fresh line of inquiry.

Furthermore, although it is clear that enunciative usage is a problem for French language learners in Morocco, it is equally crucial to take into account whether these learners also struggle with text grammar. To help French as a foreign language learners improve their written literary productions, we are forced to raise more questions and open up a number of study directions.

Naturally, these findings necessitate additional research to improve our comprehension of the occurrences we have seen.

Expanding the examination of the fundamental causes of persistent problems would be extremely pertinent, particularly with regard to irregular verb tenses and the misuse or underutilization of deictic expressions.

A method like this might investigate the probable impact of real teaching methods, how pupils view writing, or potential linguistic interference from their native tongue. The development and thorough assessment of focused educational treatments specifically meant to enhance mastery of enunciative tools that have been recognized as problematic, such modalities or reported speech, would be another viable approach. Last but not least, comparative research incorporating different educational levels or French as a foreign language learning environments, along with longitudinal methods that monitor the growth of these abilities over time, would surely improve our comprehension of these intricate problems.

REFERENCES

Benveniste, É. (2014). *The formal apparatus of enunciation. The Discourse Studies Readers. Main currents in theory and analysis*. Benjamins.

Ducrot, O. (2014). Enunciative polyphony. *Angermuller, Maingueneau, and Wodak, The Discourse Studies Reader*, 166-75.

Kerbrat-Orecchioni, C. (2009). *L'énonciation: de la subjectivité dans le langage*. Armand Colin.

Maingueneau, D. (2023). The French school of discourse analysis. In *Handbook of Political Discourse* (pp. 79–92). Edward Elgar Publishing. DOI: 10.4337/9781800373570.00014

Pollet, M. C. (2014). *L'écrit scientifique à l'aune des littéracies universitaires: approches théoriques et pratiques* (Vol. 9). Presses universitaires de Namur.

Compilation of References

A systematic review of literature reviews on artificial intelligence in education (AIED) : A roadmap to a future research agenda | Smart Learning Environments. (s. d.). Consulté 24 août 2025, à l'adresse https://link.springer.com/article/10.1186/s40561-024-00350-5

Abuhassna, H., Adnan, M. A. B. M., & Awae, F. (2024). Exploring the synergy between instructional design models and learning theories: A systematic literature review. *Contemporary Educational Technology, 16*(2), ep499. DOI: 10.30935/cedtech/14289

Adams, W. C. (2015). Conducting Semi-Structured Interviews. In Wholey, J. S., Harty, H. P., & Newcomer, K. E. (Eds.), *Handbook of Practical Program Evaluation* (pp. 492–505). Jossey-Bass., DOI: 10.1002/9781119171386.ch19

Adeusi, K. B., Jejeniwa, T. O., & Jejeniwa, T. O. (2024). *Advancing financial transparency and ethical governance: Innovative cost management and accountability in higher education and industry*. International Journal of Management & Entrepreneurship Research., DOI: 10.51594/ijmer.v6i5.1099

Adoui, A. (2023). Understanding the Role of International Higher Education in Nation Branding and Soft Power. In *International Higher Education and The Rise of Soft Power as Cultural Diplomacy*. Palgrave Macmillan., DOI: 10.1007/978-3-031-44180-6_4

Agarwal, D. V., Kaur, S., Gupta, S., Zaveri, D., Sohal, J., Shrivastav, D. A., & Shrivastav, K. (2024). Adoption of Advanced Systems for Digital Resource Management in Academic Institutions. *Evolutionary Studies in Imaginative Culture*.

Akkaya, A. (2021). Theories of educational management and leadership: By Tony Bush, California, SAGE Publications, 2020, 208 pp., £89.00 (Hardback), ISBN 978-1526432124; £26.02 (Paperback), ISBN 978-1526432131. *Educational Review, 73*(6), 800–802. DOI: 10.1080/00131911.2021.1890944

Al Khalil, M., Habash, N., & Saddiki, H. (2017). Simplification of Arabic Masterpieces for Extensive Reading : A Project Overview. *Procedia Computer Science, 117*, 192–198. DOI: 10.1016/j.procs.2017.10.109

AlAfnan, M., & MohdZuki, S. F. (2024). Malaysia's National Blockchain Roadmap: A critical discourse analysis of focus, goals, and challenges. *World Journal of English Language, 14*(5), 482. Advance online publication. DOI: 10.5430/wjel.v14n5p482

Alhafni, B., Hazim, R., Liberato, J. P., Khalil, M. A., & Habash, N. (2024). The SAMER Arabic Text Simplification Corpus. arXiv preprint arXiv:2404.18615.

Alolayyan, M. N., & Alyahya, M. S. (2023). Operational flexibility impact on hospital performance through the roles of employee engagement and management capability. *BMC Health Services Research, 23*(1), 19. Advance online publication. DOI: 10.1186/s12913-023-09029-y PMID: 36624513

Al-Subaihin, A. A., & Al-Khalifa, H. S. (2011). Al-Baseet : A proposed simplification authoring tool for the Arabic language. *2011 International Conference on Communications and Information Technology (ICCIT)*, 121-125. DOI: 10.1109/ICCITECHNOL.2011.5762661

Altbach, P. G., & Salmi, J. 2011. La voie de l'excellence académique: La création d'universités de recherche de rang mondial. Direction in Development; human development. © World Bank. https://documents.worldbank.org/curated/en/929521468340145873

Al-Thanyyan, S. S., & Azmi, A. M. (2021). Automated Text Simplification : A Survey. ACM Comput. Surv., 54(2), 43:1-43:36. DOI: 10.1145/3442695

Al-Thanyyan, S. S., & Azmi, A. M. (2023). Simplification of Arabic text : A hybrid approach integrating machine translation and transformer-based lexical model. *Journal of King Saud University. Computer and Information Sciences, 35*(8), 101662. DOI: 10.1016/j.jksuci.2023.101662

Althobaiti, M., Kruschwitz, U., & Poesio, M. (2014). AraNLP: A Java-based Library for the Processing of Arabic Text. 5.

Alvesson, M., & Sandberg, J. (2024). *Constructing research questions: Doing interesting research.* SAGE Publications Limited. DOI: 10.4135/9781529682588

Antoun, W., Baly, F., & Hajj, H. (2020). Arabert : Transformer-based model for arabic language understanding. arXiv preprint arXiv:2003.00104.

Anyanwu, A., Dawodu, S. O., Omotosho, A., Akindote, O. J., & Ewuga, S. K. (2023). Review of blockchain technology in government systems: Applications and impacts in the USA. *World Journal of Advanced Research and Reviews, 20*(3), 863–875. Advance online publication. DOI: 10.30574/wjarr.2023.20.3.2553

Athiwaratkun, B., Wilson, A. G., & Anandkumar, A. (2018). Probabilistic FastText for Multi-Sense Word Embeddings (No. arXiv:1806.02901). arXiv. /arXiv.1806. 02901DOI: 10.18653/v1/P18-1001

Avanzini, G. (1994). *Introduction to educational sciences* (3rd ed.). Privat.

Baazeem, I., Al-Khalifa, H., & Al-Salman, A. (2025). AraEyebility : Eye-Tracking Data for Arabic Text Readability. *Computation (Basel, Switzerland), 13*(5), 5. Advance online publication. DOI: 10.3390/computation13050108

Baharmand, H., Saeed, N., Comes, T., & Lauras, M. (2021). Developing a framework for designing humanitarian blockchain projects. *Computers in Industry, 131*, 103487. DOI: 10.1016/j.compind.2021.103487

Bailly, B., Demougeot-Lebel, J., & Lison, C. (2015). La formation d'enseignants universitaires nouvellement recrutés: Quelles retombées? *Revue Internationale de Pédagogie de l'Enseignement Supérieur, 31*(3). Advance online publication. https://journals.openedition.org/ripes/990. DOI: 10.4000/ripes.990

Baines, A., Medina, D., & Healy, C. (2023). *Amplify student voices: Equitable practices to build confidence in the classroom*. ASCD.

Bakos, Y., & Halaburda, H. (2022). *Overcoming the coordination problem in new marketplaces via cryptographic tokens*. SSRN Electronic Journal. DOI: 10.2139/ssrn.4028849

Balasubramanian, S., Shukla, V., Islam, N., Upadhyay, A., & Duong, L. (2023). Applying artificial intelligence in healthcare: Lessons from the COVID-19 pandemic. *International Journal of Production Research*. Advance online publication. DOI: 10.1080/00207543.2023.2263102

Baldezamo, R. C., Requillo, D. A. C., López, A. D., Manguiob, I. D., & Guray, C. B. (2024). A Systematic Review on Technology Resource Management in Education. *Asian Journal of Education and Social Studies, 50*(7), 272–285. DOI: 10.9734/ajess/2024/v50i71462

Ballandies, M. (2022). To incentivize or not: Impact of blockchain-based cryptoeconomic tokens on human information sharing behavior. *IEEE Access : Practical Innovations, Open Solutions, 10*, 74111–74130. DOI: 10.1109/ACCESS.2022.3189774

Bandura, A. (1986). *Social foundations of thought and action: A social cognitive theory*. Prentice-Hall., DOI: 10.2307/2076007

Barbereau, T. J., Smethurst, R., Papageorgiou, O., Sedlmeir, J., & Fridgen, G. (2023). Decentralised finance's timocratic governance: The distribution and exercise of tokenised voting rights. *Technology in Society, 73*, 102251. Advance online publication. DOI: 10.1016/j.techsoc.2023.102251

Barrett, K. R., Bower, B. L., & Donovan, N. J. (2007). Teaching styles of community college instructors. *American Journal of Distance Education, 21*(1), 37–49. DOI: 10.1080/08923640701298738

Bayan, T., & Banach, R. (2024). *A privacy-preserving DAO model using NFT authentication for the Punishment not Reward blockchain architecture*. ArXiv. https://doi.org//arXiv.2405.13156DOI: 10.48550

Bećirović, S. (2023). Challenges and Barriers for Effective Integration of Technologies into Teaching and Learning. In *Digital Pedagogy. SpringerBriefs in Education*. Springer., DOI: 10.1007/978-981-99-0444-0_10

Benjelloun, W. (2022). Research Policy in Morocco and the Impact on National Development. In Badran, A., Baydoun, E., & Hillman, J. R. (Eds.), *Higher Education in the Arab World: Research and Development*. Springer., DOI: 10.1007/978-3-030-80122-9_12

Benveniste, É. (2014). *The formal apparatus of enunciation. The Discourse Studies Readers. Main currents in theory and analysis*. Benjamins.

Berrichi, S., Nassiri, N., Mazroui, A., & Lakhouaja, A. (2023). Impact of Feature Vectorization Methods on Arabic Text Readability Assessment. In Farhaoui, Y., Rocha, A., Brahmia, Z., & Bhushab, B. (Eds.), *Artificial Intelligence and Smart Environment* (pp. 504–510). Springer International Publishing., DOI: 10.1007/978-3-031-26254-8_73

Bhat, S., Gijo, E. V., Antony, J., & Cross, J. (2023). Strategies for successful deployment and sustainment of Lean Six Sigma in healthcare sector in India: A multi-level perspective. *The TQM Journal, 35*(2), 414–445. Advance online publication. DOI: 10.1108/TQM-10-2021-0302

Biran, O., Brody, S., & Elhadad, N. (2011). Putting it Simply : A Context-Aware Approach to Lexical Simplification. Proceedings of the 49th Annual Meeting of the Association for Computational Linguistics: Human Language Technologies, 496-501. https://aclanthology.org/P11-2087

Black, W., Elkateb, S., Rodriguez, H., Alkhalifa, M., Vossen, P., Pease, A., & Fellbaum, C. (2006). Introducing the Arabic wordnet project. *Proceedings of the third international WordNet conference*, 295-300.

Bless, C., Denzler, A., Karras, O., & Auer, S. (2024). *A reputation system for scientific contributions based on a token economy*. In *Proceedings of the 27th International Conference on Knowledge Engineering and Knowledge Management* (pp. 41–61). Springer. DOI: 10.1007/978-3-031-72437-4_3

Boal, A. (2005). *Theater of the oppressed*. La Découverte.

Boi, B., Cirillo, F., De Santis, M., & Esposito, C. (2024). Soulbound tokens: Enabler for privacy-aware and decentralized authentication mechanism in medical data storage. *Blockchain in Healthcare Today*, *7*(2), 334. DOI: 10.30953/bhty.v7.334 PMID: 39649415

Bott, S., Rello, L., Drndarevic, B., & Saggion, H. (2012). Can Spanish Be Simpler? LexSiS: Lexical Simplification for Spanish. In M. Kay & C. Boitet (Éds.), Proceedings of COLING 2012 (p. 357-374). The COLING 2012 Organizing Committee. https://aclanthology.org/C12-1023

Boussouf, Z., Amrani, H., & Aftiss, A. (2024). Digital Transformation: What's the Reality in the Moroccan University Context? In Motahhir, S., & Bossoufi, B. (Eds.), *Digital Technologies and Applications. ICDTA 2024. Lecture Notes in Networks and Systems* (Vol. 1098). Springer., DOI: 10.1007/978-3-031-68650-4_18

Bozeman, B., & Boardman, C. (2014). Inputs, Resources and Research Collaboration. In *Research Collaboration and Team Science. SpringerBriefs in Entrepreneurship and Innovation*. Springer., DOI: 10.1007/978-3-319-06468-0_2

Branch, R. M. (2009). *Instructional Design : The ADDIE Approach*. Springer US., DOI: 10.1007/978-0-387-09506-6

Braun, V., & Clarke, V. (2019). Reflecting on reflexive thematic analysis. *Qualitative Research in Sport, Exercise and Health*, *11*(4), 589–597. DOI: 10.1080/2159676X.2019.1628806

Bush, T. (2008). *Leadership and management development in education*. SAGE Publications Ltd., DOI: 10.4135/9781446213605

Caeymaex, F., Wenger, C., De Heusch, F., & Lafleur, J. M. (2023). Ethics Ready"? *Governing Research Through Informed Consent Procedures. International Journal of Qualitative Methods*, *22*, 16094069231165718.

California State University. (2017). *CSU Quality Online Learning & Teaching (QOLT)*. https://dspace.calstate.edu/handle/10211.3/161320

Cannavale, C., Esempio Tammaro, A., Leone, D., & Schiavone, F. (2022). Innovation adoption in inter-organizational healthcare networks – the role of artificial intelligence. In *European Journal of Innovation Management* (Vol. 25, Issue 6). DOI: 10.1108/EJIM-08-2021-0378

Carter, M. W., & Busby, C. R. (2023). How can operational research make a real difference in healthcare? Challenges of implementation. *European Journal of Operational Research, 306*(3), 1059–1068. Advance online publication. DOI: 10.1016/j.ejor.2022.04.022

Choetkiertikul, M., Puengmongkolchaikit, A., Chandra, P., Ragkhitwetsagul, C., Maipradit, R., Hata, H., Sunetnanta, T., & Matsumoto, K. (2023). Studying the association between Gitcoin's issues and resolving outcomes. *Journal of Systems and Software, 206*, 111835. DOI: 10.1016/j.jss.2023.111835

Clark, J. L. D., & Clifford, R. T. (1988). THE FSI/ILR/ACTFL PROFICIENCY SCALES AND TESTING TECHNIQUES : Development, Current Status, and Needed Research. [JSTOR.]. *Studies in Second Language Acquisition, 10*(2), 129–147. DOI: 10.1017/S0272263100007270

Coristine, S., Russo, S., Fitzmorris, R., Beninato, P., & Rivolta, G. (2022, April). *The importance of student-teacher relationships*. In *Classroom Practice in 2022*. https://ecampusontario.pressbooks.pub/educ5202/chapter/the-importance-of-student-teacher-relationships/

Creswell, J. W., & Poth, C. N. (2018). *Qualitative inquiry and research design: Choosing among five approaches* (4th ed.). Sage.

Crowe, S., & Utley, M. (2022). Praxis in healthcare OR: An empirical behavioural OR study. *The Journal of the Operational Research Society, 73*(7), 1444–1456. Advance online publication. DOI: 10.1080/01605682.2021.1919036

Czerniewicz, L., Agherdien, N., Badenhorst, J., Belluigi, D., Chambers, T., Chili, M., de Villiers, M., Felix, A., Gachago, D., Gokhale, C., Ivala, E., Kramm, N., Madiba, M., Mistri, G., Mgqwashu, E., Pallitt, N., Prinsloo, P., Solomon, K., Strydom, S., & Wissing, G. (2020). A Wake-Up Call: Equity, Inequality and Covid-19 Emergency Remote Teaching and Learning. *Postdigit Sci Educ, 2*(3), 946–967. DOI: 10.1007/s42438-020-00187-4 PMID: 40477091

De Belder, J., & Moens, M.-F. (2010). Text simplification for children. Prroceedings of the SIGIR workshop on accessible search systems, 19-26.

de Peretti, A. (2003). *Training professional teachers*. ESF.

Deci, E. L., & Ryan, R. M. (1985). *Intrinsic motivation and self-determination in human behavior*. Springer., DOI: 10.1007/978-1-4899-2271-7

Deci, E., Vallerand, R., Pelletier, L., & Ryan, R. (1991). Motivation and education: The self-determination perspective. *Educational Psychologist*, *26*(3), 325–346. DOI: 10.1207/s15326985ep2603&4_6

Deem, R. (2001). Globalisation, New Managerialism, Academic Capitalism and Entrepreneurialism in Universities: Is the Local Dimension Still Important? *Comparative Education*, *37*(1), 7–20. https://www.jstor.org/stable/3099730. DOI: 10.1080/03050060020020408

Deem, R., & Brehony, K. J. (2005). Management as ideology: The case of 'new managerialism' in higher education. *Oxford Review of Education*, *31*(2), 217–235. DOI: 10.1080/03054980500117827

Denzin, N. K., & Lincoln, Y. S. (Eds.). (2017). *The SAGE handbook of qualitative research* (5th ed.). SAGE Publications.

Detwal, P. K., Agrawal, R., Samadhiya, A., Kumar, A., & Garza-Reyes, J. A. (2024). Revolutionizing healthcare organizations with Operational Excellence and Healthcare 4.0: a systematic review of the state-of-the-art literature. In *International Journal of Lean Six Sigma* (Vol. 15, Issue 1). DOI: 10.1108/IJLSS-04-2023-0061

Dion, H., & Evans, M. (2024). Strategic frameworks for sustainability and corporate governance in healthcare facilities; approaches to energy-efficient hospital management. *Benchmarking*, *31*(2), 353–390. Advance online publication. DOI: 10.1108/BIJ-04-2022-0219

Directorate of Financial Studies and Forecasting (DEPF). (2022). *Report on linguistic competencies in Moroccan secondary education*. Ministry of Economy and Finance.

Diwas Singh, K. C., Scholtes, S., & Terwiesch, C. (2020). Empirical research in healthcare operations: Past research, present understanding, and future opportunities. *Manufacturing & Service Operations Management*, *22*(1), 73–83. Advance online publication. DOI: 10.1287/msom.2019.0826

Dodge, J. R., Youles, B., Caldararo, J., Sears, E. D., Caverly, T. J., Michael Ho, P., Shimada, S. L., Kaboli, P., Albright, K., Robinson, S. A., McNeal, D. M., Damschroder, L., Saini, S. D., & Adams, M. A. (2023). Engaging Operational Partners Is Critical for Successful Implementation of Research Products: A Coincidence Analysis of Access-Related Projects in the Veterans Affairs Healthcare System. *Journal of General Internal Medicine*, *38*(S3), 923–930. Advance online publication. DOI: 10.1007/s11606-023-08115-5 PMID: 37340262

Dodson, R. A. S. (2023). Community College Management Strategies for Balancing the Budgets (Doctoral dissertation, Walden University).

Domenicale, I., Viano, C., & Schifanella, C. (2024). Blockchain for local communities: An exploratory review of token economy aspects. *Frontiers in Blockchain*, *7*, 1426802. DOI: 10.3389/fbloc.2024.1426802

Dong, Y. (2023). Teaching quality monitoring and evaluation in higher education through a big data analysis. *International Journal of Emerging Technologies in Learning*, *18*(08), 61–78. DOI: 10.3991/ijet.v18i08.39247

Ducrée, J., Etzrodt, M., Bartling, S., Walshe, R., Wittek, N., & Kogias, D. (2021). Unchaining collective intelligence for science, research, and technology development by blockchain-boosted community participation. *Frontiers in Blockchain*, *4*, 631648. DOI: 10.3389/fbloc.2021.631648

Ducrot, O. (2014). Enunciative polyphony. *Angermuller, Maingueneau, and Wodak, The Discourse Studies Reader*, 166-75.

Dwikat, S. Y., Arshad, D., & Mohd Shariff, M. N. (2022). The influence of systematic strategic planning and strategic business innovation on the sustainable performance of manufacturing SMEs: The case of Palestine. *Sustainability (Basel)*, *14*(20), 13388. DOI: 10.3390/su142013388

Dzando, G., Salifu, S., Donyi, A. B., Akpeke, H., Kumah, A., Dordunu, R., & Nonoh, E. A. (2022). Healthcare in Ghana amidst the coronavirus pandemic: A narrative literature review. In *Journal of Public Health Research* (Vol. 11, Issue 1). DOI: 10.4081/jphr.2021.2448

Elansari, H., & Loulid, M. (2023). STUDY OF THE DIGITAL TRANSFORMATION OF HIGHER EDUCATION ESTABLISHMENTS: THE CASE OF THE MOROCCAN UNIVERSITY. *MINAR International Journal of Applied Sciences and Technology*, *05*(04), 136–155. DOI: 10.47832/2717-8234.17.10

Elmadani, K. N., Habash, N., & Taha-Thomure, H. (2025). A Large and Balanced Corpus for Fine-grained Arabic Readability Assessment (No. arXiv:2502.13520). arXiv. /arXiv.2502.13520DOI: 10.18653/v1/2025.findings-acl.842

Errihani, M. (2023). Education and Language Reforms. In *Language, Power, and the Economics of Education in Morocco. Language Policy* (Vol. 35). Springer., DOI: 10.1007/978-3-031-51594-1_4

Esmaeilzadeh, P., & Mirzaei, T. (2023). Role of incentives in the use of blockchain-based platforms for sharing sensitive health data. *Journal of Medical Internet Research, 25*, e41805. Advance online publication. DOI: 10.2196/41805 PMID: 37594783

Evans, N. G. (2024). *Strategic management for tourism, hospitality and events.* Routledge. DOI: 10.4324/9781003318613

Even, A. M. (2024). Project Management and Education: Improving Learning and Student Success. [IJCDLM]. *International Journal of Curriculum Development and Learning Measurement, 5*(1), 1–27. DOI: 10.4018/IJCDLM.353926

E-xcellence. (2012). *Quality assessment for e-learning: A benchmarking approach.* https://e-xcellencelabel.eadtu.eu/images/documents/Excellence_manual_full.pdf

Farahani, R. Z., Ruiz, R., & Van Wassenhove, L. N. (2023). Introduction to the special issue on the role of operational research in future epidemics/ pandemics. In *European Journal of Operational Research* (Vol. 304, Issue 1). DOI: 10.1016/j.ejor.2022.07.019

Fekete, D., & Kiss, A. (2023). Toward building smart contract-based higher education systems using zero-knowledge Ethereum virtual machine. *Electronics (Basel), 12*(3), 664. DOI: 10.3390/electronics12030664

Fernandes, F., & Werner, C. (2023). *Towards a blockchain-based software engineering education.* ArXiv. /arXiv.2304.04549DOI: 10.5753/cblockchain.2023.230488

Finn, E., & Wylie, R. (2021). Collaborative imagination: A methodological approach. *Futures, 132*, 102788. DOI: 10.1016/j.futures.2021.102788

Fitzpatrick, G., Decroo, T., Draguez, B., Crestani, R., Ronsse, A., Van den Bergh, R., & Van Herp, M. (2017). Operational research during the Ebola emergency. *Emerging Infectious Diseases, 23*(7), 1057–1062. Advance online publication. DOI: 10.3201/eid2307.161389 PMID: 28628463

Freire, P. (1970). *Pedagogy of the oppressed.* Maspero. (Re-ed. 2021, Éditions La Découverte)

Fusco, F., Marsilio, M., & Guglielmetti, C. (2023). Co-creation in healthcare: framing the outcomes and their determinants. In *Journal of Service Management* (Vol. 34, Issue 6). DOI: 10.1108/JOSM-06-2021-0212

Garrido, J. M. (2023). *Digital tokens: A legal perspective*. IMF Working Papers. DOI: 10.5089/9798400250149.001

Gioia, D. (2021). A systematic methodology for doing qualitative research. *The Journal of Applied Behavioral Science*, *57*(1), 20–29. DOI: 10.1177/0021886320982715

Golam, M., Tuli, E. A., Alief, R. N., Kim, D. S., & Lee, J. M. (2024). Meta-learning: A digital learning management framework using blockchain for metaverses. *IEEE Access : Practical Innovations, Open Solutions*, *12*, 92774–92786. DOI: 10.1109/ACCESS.2024.3408878

Grosjean, B. (2007). *Theater as a pedagogical tool for language classes*. Hachette FLE.

Gunawan, I., Lutfiani, N., Aini, Q., Suryaman, F. M., & Sunarya, A. (2021). Smart contract innovation and blockchain-based tokenization in higher education. *Journal of Education Technology*, *5*(4), 256–264. DOI: 10.23887/jet.v5i4.40665

Hallam, C. R. A., & Contreras, C. (2018). Lean healthcare: scale, scope and sustainability. In *International Journal of Health Care Quality Assurance* (Vol. 31, Issue 7). DOI: 10.1108/IJHCQA-02-2017-0023

Hallwright, J., & Carnaby, E. (2019). Complexities of implementation: Oxfam Australia's experience in piloting blockchain. *Frontiers in Blockchain*, *2*, 10. DOI: 10.3389/fbloc.2019.00010

Hassan I Middle School Teaching Team. (2024, October 13). *Pedagogical booklet for the "Innovative Theater" project* [Internal document].

Heizelman, S., & Bittner, K. (2024). *Empowering engineering students: Blockchain learning tokens for assessing ABET outcomes*. ASEE., DOI: 10.18260/1-2--47245

Horn, C., Manduca, C., & Kauchak, D. (2014). Learning a lexical simplifier using wikipedia. *Proceedings of the 52nd Annual Meeting of the Association for Computational Linguistics* (Volume 2: Short Papers), 458-463.

Hrivnák, M., Moritz, P., Melichová, K., Roháčiková, O., & Pospišová, L. (2021). Designing the participation on local development planning: From literature review to adaptive framework for practice. *Societies (Basel, Switzerland)*, *11*(1), 19. DOI: 10.3390/soc11010019

Hua, X. (2021). Construction quality evaluation of teacher development schools based on improved artificial neural network. *International Journal of Emerging Technologies in Learning*, *16*(08), 205–220. DOI: 10.3991/ijet.v16i08.22129

Hubinger, E., Denison, C., Mu, J., Lambert, M., Tong, M., MacDiarmid, M., . . . Perez, E. (2024). Sleeper agents: Training deceptive llms that persist through safety training. arXiv preprint arXiv:2401.05566.

Hunt, L. A. (2024). Work, Legacy, and a Personal Reflection. In *Work and the Older Person* (pp. 147–150). Routledge.

Hussain, M., Ajmal, M., Subramanian, G., Khan, M., & Anas, S. (2023). Challenges of big data analytics for sustainable supply chains in healthcare – a resource-based view. *Benchmarking*. Advance online publication. DOI: 10.1108/BIJ-06-2022-0390

Iheukwumere-Esotu, L. O., & Yunusa-Kaltungo, A. (2021). Knowledge management and experience transfer in major maintenance activities: A practitioner's perspective. *Sustainability (Basel)*, *14*(1), 52. DOI: 10.3390/su14010052

Ilangakoon, T. S., Weerabahu, S. K., Samaranayake, P., & Wickramarachchi, R. (2022). Adoption of Industry 4.0 and lean concepts in hospitals for healthcare operational performance improvement. *International Journal of Productivity and Performance Management*, *71*(6), 2188–2213. Advance online publication. DOI: 10.1108/IJPPM-12-2020-0654

Interview with a teacher from Hassan I Middle School. (2024, November 12). Morocco. [Unpublished primary source]

Ivanova, E. (2023). Adaptive Leadership in Times of Uncertainty: Embracing Change and Fostering Collective Growth. *Abbottabad University Journal of Business and Management Sciences*, *1*(02), 133–138.

Jha, A., Athanerey, A., & Kumar, A. (2022). Role and challenges of internet of things and informatics in Healthcare research. In *Health and Technology* (Vol. 12, Issue 4). DOI: 10.1007/s12553-022-00661-y

Johnson, A. M., Jacovina, M. E., Russell, D. G., & Soto, C. M. (2016). Challenges and Solutions when Using Technologies in the Classroom. In *Routledge eBooks* (pp. 13–30). DOI: 10.4324/9781315647500-2

Joshi, S., Sharma, M., Das, R. P., Rosak-Szyrocka, J., Żywiołek, J., Muduli, K., & Prasad, M. (2022). Modeling Conceptual Framework for Implementing Barriers of AI in Public Healthcare for Improving Operational Excellence: Experiences from Developing Countries. *Sustainability (Basel)*, *14*(18), 11698. Advance online publication. DOI: 10.3390/su141811698

Kakalou, E., Riza, E., Chalikias, M., Voudouri, N., Vetsika, A., Tsiamis, C., Choursoglou, S., Terzidis, A., Karamagioli, E., Antypas, T., & Pikoulis, E. (2018). Demographic and clinical characteristics of refugees seeking primary healthcare services in Greece in the period 2015-2016: A descriptive study. *International Health, 10*(6), 421–429. Advance online publication. DOI: 10.1093/inthealth/ihy042 PMID: 29992276

Kamila, J. S., & Marzuq, M. F. (2024). Asana and Trello: A comparative assessment of project management capabilities. *JOIV: International Journal on Informatics Visualization, 8*(1), 207–212. DOI: 10.62527/joiv.8.1.2595

Karat, G., & Kannimoola, J. M. (2023). *Revolutionizing education: Blockchain-based incentive systems*. ICICCS, 1370–1375. DOI: 10.1109/ICICCS56967.2023.10142632

Kayyali, M. (2025b). Strategic Planning in Higher Education: Frameworks and Best Practices. In Planning Tools for Policy, Leadership, and Management of Education Systems (pp. 251-286). IGI Global.

Kayyali, M. (2026). Research Methodology: Qualitative, Quantitative, and Mixed Methods. In Navigating Theory, Methodology, and Researcher Development in Doctoral Research (pp. 93-122). IGI Global Scientific Publishing.

Kayyali, M. (2025a). A Practical Guide to Strategic Planning in Higher Education. In *Building Organizational Capacity and Strategic Management in Academia* (pp. 1–34). IGI Global Scientific Publishing.

Kenny, J. (2002). Managing innovation in educational institutions. *Australasian Journal of Educational Technology, 18*(3). Advance online publication. DOI: 10.14742/ajet.1765

Kerbrat-Orecchioni, C. (2009). *L'énonciation: de la subjectivité dans le langage*. Armand Colin.

Kerroum, K., Khiat, A., Bahnasse, A., Aoula, E., & Khiat, Y. (2020). The proposal of an agile model for the digital transformation of the University Hassan II of Casablanca 4.0. *Procedia Computer Science, 175*, 403–410. DOI: 10.1016/j.procs.2020.07.057 PMID: 32834879

Keskinocak, P., & Savva, N. (2020). A review of the healthcare-management (modeling) literature published in manufacturing & service operations management. *Manufacturing & Service Operations Management, 22*(1), 59–72. Advance online publication. DOI: 10.1287/msom.2019.0817

Khallaf, N., & Sharoff, S. (2022, avril 20). Towards Arabic Sentence Simplification via Classification and Generative Approaches. arXiv.Org. https://arxiv.org/abs/2204.09292v1 DOI: 10.18653/v1/2022.wanlp-1.5

Khandker, S. R., Lavy, V., & Filmer, D. (1994). Schooling and cognitive achievements of children in Morocco: Can the government improve outcomes? World Bank Discussion Papers. The World Bank. DOI: 10.1596/0-8213-3046-2

Khan, M. F. F., Viet, P. D., & Sakamura, K. (2022). *Blockchain-based incentive management system for educational organizations*. ICECCME., DOI: 10.1109/ICECCME55909.2022.9988717

Kosmarski, A., & Gordiychuk, N. (2020). Token-curated registry in a scholarly journal. *Learned Publishing*, *33*(4), 370–379. DOI: 10.1002/leap.1302

Kraus, S., Schiavone, F., Pluzhnikova, A., & Invernizzi, A. C. (2021). Digital transformation in healthcare: Analyzing the current state-of-research. *Journal of Business Research*, *123*, 557–567. Advance online publication. DOI: 10.1016/j.jbusres.2020.10.030

Küng, R., & Giaglis, G. M. (2025). *Value distribution in OSS: The role of DAOs*. ICEDEG, 160–168. DOI: 10.1109/ICEDEG65568.2025.11081577

Lamberty, R., Poddey, A., Galindo, D., de Waard, D., Koelbel, T., & Kirste, D. (2020). Efficiency in digital economies: A primer on tokenomics. *Digital Economics Review*, *4*(2), 55–73.

Lamé, G., Crowe, S., & Barclay, M. (2022). "What's the evidence?"—Towards more empirical evaluations of the impact of OR interventions in healthcare. *Health Systems (Basingstoke, England)*, *11*(1), 59–67. Advance online publication. DOI: 10.1080/20476965.2020.1857663 PMID: 35127059

Lamé, G., Crowe, S., Komashie, A., & Royston, G. (2023). Joining forces: The value of design partnering with operational research to improve healthcare delivery. *Design Science*, *9*, e4. Advance online publication. DOI: 10.1017/dsj.2023.2

Layton, M. C., Ostermiller, S. J., & Kynaston, D. J. (2025). *Agile project management for dummies*. John Wiley & Sons.

Lee, J., Moroso, M., & Mackey, T. (2023). Unblocking recognition: A token system for academic contribution. *Frontiers in Blockchain*, *6*, 1136641. DOI: 10.3389/fbloc.2023.1136641

Lehman, B. R. (2024). The Question of Questioning: Studying the Sensemaking Process in Ambiguous Scenarios (Doctoral dissertation, Michigan Technological University).

Leite, H., Lindsay, C., & Kumar, M. (2021). COVID-19 outbreak: Implications on healthcare operations. *The TQM Journal*, *33*(1), 247–256. Advance online publication. DOI: 10.1108/TQM-05-2020-0111

Leone, D., Schiavone, F., Appio, F. P., & Chiao, B. (2021). How does artificial intelligence enable and enhance value co-creation in industrial markets? An exploratory case study in the healthcare ecosystem. *Journal of Business Research*, *129*, 849–859. Advance online publication. DOI: 10.1016/j.jbusres.2020.11.008

Lesavre, L., Varin, P., & Yaga, D. (2021). Blockchain networks: Token design and management overview. *NISTIR*, *8301*. Advance online publication. DOI: 10.6028/NIST.IR.8301

Lim, J. O., & Kamarudin, D. (2023). NFTs in combatting certificate fraudulence. *IJEBT*, *1*(1), 1–15. DOI: 10.59021/ijebt.v1i1.13

Lo, C. (2023). Checking In" within classroom and digital spaces to build communities of support. Developing trauma-informed teachers: Creating classrooms that foster equity, resilience, and assetbased approaches: Research findings from the field, 25-47.

Luyts, J., Burnay, N., Piguet, E., Fall, A., Mballo, I., De Longueville, F., & Henry, S. (2024). Structured timeline mapping as a data collection methodology: A new perspective for research on environmental adaptation. *Climate and Development*, , 1–11.

Mahjoob, A., Alfadhli, Y., & Omachonu, V. (2023). Healthcare Waste and Sustainability: Implications for a Circular Economy. *Sustainability (Basel)*, *15*(10), 7788. Advance online publication. DOI: 10.3390/su15107788

Mahraz, M. (2018, July). Critical Success Factors for ERP Implementations: A Moroccan Case Study. In *2nd European International Conference on Industrial Engineering and Operations Management*, https://doi.org/DOI: 10.46254/EU02.20180204

Maingueneau, D. (2023). The French school of discourse analysis. In *Handbook of Political Discourse* (pp. 79–92). Edward Elgar Publishing. DOI: 10.4337/9781800373570.00014

Maister, D. H., Galford, R., & Green, C. (2021). *The trusted advisor*. Free Press.

Maki, P. L. (2023). *Assessing for learning: Building a sustainable commitment across the institution*. Routledge. DOI: 10.4324/9781003443056

Maley, C. H. (2024). *How Not to Fail at Projects: Stopping the Project Management Insanity Spiral*. CRC Press. DOI: 10.1201/9781003471691

Mamlouk, Z., & Dhaouadi, N. (2007). E-formation et développement des compétences. *La Revue des Sciences de Gestion, 226–227*(4), 173–182. DOI: 10.1051/larsg:2007046

Markham, A. N., Tiidenberg, K., & Herman, A. (2018). Ethics as Methods: Doing Ethics in the Era of Big Data Research—Introduction. *Social Media + Society, 4*(3). (Original work published 2018)DOI: 10.1177/2056305118784502

Marsilio, M., Pisarra, M., Rubio, K., & Shortell, S. (2022). Lean adoption, implementation, and outcomes in public hospitals: Benchmarking the US and Italy health systems. *BMC Health Services Research, 22*(1), 122. Advance online publication. DOI: 10.1186/s12913-022-07473-w PMID: 35090455

Matters, Q. (2017). *Specific review standards from the QM Higher Education Rubric* (7th ed.)., https://www.qualitymatters.org/sites/default/files/PDFs/StandardsfromtheQMHigherEducationRubric.pdf

Mazzardo, E. (2020). *A token economy for school performance optimization*. AARN. DOI: 10.2139/ssrn.3696544

McDermott, O., Antony, J., & Douglas, J. (2021). Exploring the use of operational excellence methodologies in the era of COVID-19: Perspectives from leading academics and practitioners. *The TQM Journal, 33*(8), 1647–1665. Advance online publication. DOI: 10.1108/TQM-01-2021-0016

Meirieu, P. (1996). *Pedagogy between theory and practice*. ESF.

Mikolov, T., Chen, K., Corrado, G., & Dean, J. (2013). Efficient estimation of word representations in vector space. arXiv preprint arXiv:1301.3781.

Miles, M. B., Huberman, A. M., & Saldana, J. (2018). *Qualitative data analysis: A methods sourcebook* (4th ed.). SAGE Publications, Inc.

Minh Dang, L., Piran, M. J., Han, D., Min, K., & Moon, H. (2019). A survey on internet of things and cloud computing for healthcare. *Electronics (Basel), 8*(7), 768. Advance online publication. DOI: 10.3390/electronics8070768

Ministry of National Education. (2020). *National survey on family language practices*. MEN.

Mohammad, A., & Vargas, S. (2022). Challenges of using blockchain in education. *Applied Sciences (Basel, Switzerland), 12*(13), 6380. Advance online publication. DOI: 10.3390/app12136380

Molière. (1666). *The doctor in spite of himself.* Jean Ribou. (Public domainedition, Gallica BnF)

Moore, J. C., & Shelton, K. (2013). Social and student engagement and support: The Sloan-C quality scorecard for the administration of online programs. *Online Learning : the Official Journal of the Online Learning Consortium, 17*(1). Advance online publication. DOI: 10.24059/olj.v17i1.342

Morales-Alarcón, C. H., Bodero-Poveda, E., Villa-Yánez, H. M., & Buñay-Guisñan, P. A. (2024). *Blockchain & peer review.* Publications., DOI: 10.3390/publications12040040

Moroccan Ministry of National Education. (2021). *National French curriculum for middle school.* MEN.

Morrissey, D. (2023). The Pull of Opposing Forces: An Inquiry Into the Ethical Dimensions of an Emergent Research-Based Theatre Project. *Qualitative Inquiry, 29*(2), 374–382. DOI: 10.1177/10778004221097059

Mustafa, G., Rafiq, W., Jhamat, N., Arshad, Z., & Rana, F. A. (2024). *Blockchain governance models in e-government.* IJLM., DOI: 10.1108/IJLMA-08-2023-0172

Mwanza, J., Telukdarie, A., & Igusa, T. (2023). Impact of industry 4.0 on healthcare systems of low- and middle- income countries: a systematic review. In *Health and Technology* (Vol. 13, Issue 1). DOI: 10.1007/s12553-022-00714-2

Nair, D., Thekkur, P., Fernando, M., Kumar, A. M. V., Satyanarayana, S., Chandraratne, N., Chandrasiri, A., Attygalle, D. E., Higashi, H., Bandara, J., Berger, S. D., & Harries, A. D. (2023). Outcomes and Challenges in Noncommunicable Disease Care Provision in Health Facilities Supported by Primary Health Care System Strengthening Project in Sri Lanka: A Mixed-Methods Study. *Healthcare (Basel), 11*(2), 202. Advance online publication. DOI: 10.3390/healthcare11020202 PMID: 36673570

Nassiri, N., Cavalli-Sforza, V., & Lakhouaja, A. (2023). Approaches, Methods, and Resources for Assessing the Readability of Arabic Texts. ACM Transactions on Asian and Low-Resource Language Information Processing, 22(4), 95:1-95:30. DOI: 10.1145/3571510

Nassiri, N., Lakhouaja, A., & Cavalli-Sforza, V. (2018a). Arabic Readability Assessment for Foreign Language Learners. In Silberztein, M., Atigui, F., Kornyshova, E., Métais, E., & Meziane, F. (Eds.), *Natural Language Processing and Information Systems* (pp. 480–488). Springer International Publishing. DOI: 10.1007/978-3-319-91947-8_49

Nassiri, N., Lakhouaja, A., & Cavalli-Sforza, V. (2018b). Modern Standard Arabic Readability Prediction. In Lachkar, A., Bouzoubaa, K., Mazroui, A., Hamdani, A., & Lekhouaja, A. (Eds.), *Arabic Language Processing : From Theory to Practice* (pp. 120–133). Springer International Publishing.

Nassiri, N., Lakhouaja, A., & Cavalli-Sforza, V. (2022a). Arabic L2 readability assessment : Dimensionality reduction study. *Journal of King Saud University. Computer and Information Sciences*, *34*(6, 6, Part B), 3789–3799. DOI: 10.1016/j.jksuci.2020.12.021

Nassiri, N., Lakhouaja, A., & Cavalli-Sforza, V. (2022b). Combining Classical and Non-classical Features to Improve Readability Measures for Arabic First Language Texts. In Kacprzyk, J., Balas, V. E., & Ezziyyani, M. (Eds.), *Advanced Intelligent Systems for Sustainable Development (AI2SD'2020)* (pp. 463–470). Springer International Publishing., DOI: 10.1007/978-3-030-90633-7_39

Nguyen, P.-B.-T., Degrave, P., Van Steendam, E., & Sercu, L. (2023). *Self-determination in EMI*. IJ Educational Research Open., DOI: 10.1016/j.ijedro.2023.100295

Ning, G. (2021). Evaluation criteria for quality education of physical education lessons based on logical analysis. *International Journal of Emerging Technologies in Learning*, *16*(21), 87–99. DOI: 10.3991/ijet.v16i21.26867

Niu, M. (2023). Quality evaluation of innovation and entrepreneurship education based on modelling of students' behavior sequences. *International Journal of Emerging Technologies in Learning*, *18*(09), 17–31. DOI: 10.3991/ijet.v18i09.40227

OECD. (2014). *Promoting Research Excellence: New Approaches to Funding*. OECD Publishing., DOI: 10.1787/9789264207462-

Olssen, M., & Peters, M. A. (2005). Neoliberalism, higher education and the knowledge economy: From the free market to knowledge capitalism. *Journal of Education Policy*, *20*(3), 313–345. DOI: 10.1080/02680930500108718

Oshchepkova, T. (2023). First Steps to Research: Typical Mistakes That Undergraduate Students Make When They Approach Academic Writing. *Journal of Curriculum and Teaching*, *12*(1), 212. DOI: 10.5430/jct.v12n1p212

Paddeu, D., Lyons, G., Chatterjee, K., & Calvert, T. (2024). Practitioner views on transport planning's evolution–A Sisyphean task still ahead? *Transport Policy, 156*, 89–100. DOI: 10.1016/j.tranpol.2024.07.015

Paetzold, G. H., & Specia, L. (2017). A Survey on Lexical Simplification. *Journal of Artificial Intelligence Research, 60*, 549–593. DOI: 10.1613/jair.5526

Palmer, R., Fulop, N. J., & Utley, M. (2018). A systematic literature review of operational research methods for modelling patient flow and outcomes within community healthcare and other settings. *Health Systems (Basingstoke, England), 7*(1), 29–50. Advance online publication. DOI: 10.1057/s41306-017-0024-9 PMID: 31214337

Panagiotidis, P. (2022). Blockchain in language learning. *European Journal of Education, 5*(1), 66–82. DOI: 10.26417/443gjm83

Pasha, A., Al-Badrashiny, M., Diab, M. T., El Kholy, A., Eskander, R., Habash, N., Pooleery, M., Rambow, O., & Roth, R. (2014). MADAMIRA: A Fast, Comprehensive Tool for Morphological Analysis and Disambiguation of Arabic. *LREC, 14*, 1094–1101.

Pasquini, R. (2020). *Quadratic funding and matching funds*. Working Paper.

Patel, J., More, S., Sohani, P., Bedarkar, S., Dinesh, K. K., Sharma, D., Dhir, S., Sushil, S., & Ghosh, R. S. (2024). Reshaping the equitable and inclusive access to healthcare: A qualitative study. *Clinical Epidemiology and Global Health, 26*, 101544. Advance online publication. DOI: 10.1016/j.cegh.2024.101544 PMID: 38707587

Patton, M. Q. (2022). Impact-driven qualitative research and evaluation. The SAGE handbook of qualitative research design, 2, 1165-1180. DOI: 10.4135/9781529770278.n71

Pauwels, K., & Aksehirli, Z. (2025). Big data analytics democratized with clean collaboration and customer privacy choice. *Journal of Business Research, 188*, 115112. DOI: 10.1016/j.jbusres.2024.115112

Pearson, M., Monks, T., Gibson, A., Allen, M., Komashie, A., Fordyce, A., Harris-Golesworthy, F., Pitt, M. A., Brailsford, S., & Stein, K. (2013). Involving patients and the public in healthcare operational research-The challenges and opportunities. *Operations Research for Health Care, 2*(4), 86–89. Advance online publication. DOI: 10.1016/j.orhc.2013.09.001

Pereira, L. M., Crespo, G. O., Amon, D. J., Badhe, R., Bandeira, S., Bengtsson, F., & Zhou, W. (2023). The living infinite: Envisioning futures for transformed human-nature relationships on the high seas. *Marine Policy, 153*, 105644. DOI: 10.1016/j.marpol.2023.105644

Pericàs-Gornals, R.. (2024). Rejectable soulbound tokens in credentials. *Annales des Télécommunications*, *79*, 843–855. DOI: 10.1007/s12243-024-01032-6

Perrenoud, P. (1994). *Teacher training between theory and practice*. L'Harmattan.

Pignarre, R. (1979). *History of theater*. Presses Universitaires de France.

Pirro, N. J. (2024). *Where Are You Headed? Get There.: A Guide to Overcoming Self-doubt for Individuals Seeking Personal and Professional Growth*. Pyrrhic Press.

Pollet, M. C. (2014). *L'écrit scientifique à l'aune des littéracies universitaires: approches théoriques et pratiques* (Vol. 9). Presses universitaires de Namur.

Pradhan, B., Bhattacharyya, S., & Pal, K. (2021). IoT-Based Applications in Healthcare Devices. In *Journal of Healthcare Engineering* (Vol. 2021). DOI: 10.1155/2021/6632599

Preethi Krishna, P., & Sharada, A. (2020). Word Embeddings—Skip Gram Model. In Gunjan, V. K., Garcia Diaz, V., Cardona, M., Solanki, V. K., & Sunitha, K. V. N. (Eds.), *ICICCT 2019 – System Reliability, Quality Control, Safety, Maintenance and Management* (pp. 133–139). Springer., DOI: 10.1007/978-981-13-8461-5_15

Project Management Institute (Ed.). (2008). *A guide to the project management body of knowledge (PMBOK guide)* (4th ed.). Project Management Institute, Inc.

Project Management Institute. (2021). *A guide to the project management body of knowledge (PMBOK® Guide)* (7th ed.). Project Management Institute.

Qiang, J., Li, Y., Zhu, Y., Yuan, Y., Shi, Y., & Wu, X. (2021). LSBert : Lexical simplification based on BERT. *IEEE/ACM Transactions on Audio, Speech, and Language Processing*, *29*, 3064–3076. DOI: 10.1109/TASLP.2021.3111589

R'boul, H. (2024). Higher Education: Recent Developments. In: Education in Morocco. Palgrave Macmillan, Cham. DOI: 10.1007/978-3-031-72414-5_3

Ra, G., Seo, D., Bhuiyan, M. Z. A., & Lee, I. (2020). Anonymous protocol for privacy in blockchain. *Electronics (Basel)*, *9*(8), 1183. DOI: 10.3390/electronics9081183

Rais, A., & Vianaa, A. (2011). Operations research in healthcare: A survey. *International Transactions in Operational Research*, *18*(1), 1–31. Advance online publication. DOI: 10.1111/j.1475-3995.2010.00767.x

Ramalingam, S., Subramanian, M., Sreevallabha Reddy, A., Tarakaramu, N., Ijaz Khan, M., Abdullaev, S., & Dhahbi, S. (2024). Exploring business intelligence applications in the healthcare industry: A comprehensive analysis. *Egyptian Informatics Journal*, *25*, 100438. Advance online publication. DOI: 10.1016/j.eij.2024.100438

Ramos, R. E., Balatero, M. F. T., Cagape, E. M., & Guray, C. B. (2024). Strategies for Resource Management in the Department of Education: A Systematic Review of Efficiency and Effectiveness. *Asian Journal of Education and Social Studies, 50*(8), 100–116. DOI: 10.9734/ajess/2024/v50i81510

Ryngaert, J.-P. (1998). *Acting and performing: Drama practices and training.* Armand Colin.

Sachdeva, R., Williams, T., & Quigley, J. (2007). Mixing methodologies to enhance the implementation of healthcare operational research. *The Journal of the Operational Research Society, 58*(2), 159–167. Advance online publication. DOI: 10.1057/palgrave.jors.2602293

Salman, S. (2023). Harmony in Purpose: The Key Role of Organizational Values in Nonprofit Collaboration and Cross-Sector Partnerships.

Samala, A.. (2024). Blockchain in education: Opportunities and challenges. *IJIMT, 18*(1), 20–42. DOI: 10.3991/ijim.v18i01.46307

Saraswat, D., Bhattacharya, P., Verma, A., Prasad, V. K., Tanwar, S., Sharma, G., Bokoro, P. N., & Sharma, R. (2022). Explainable AI for Healthcare 5.0: Opportunities and Challenges. In *IEEE Access* (Vol. 10). DOI: 10.1109/ACCESS.2022.3197671

Sarno, D., Pels, J., Polese, F., Sörhammar, D., Tronvoll, B., & Vargo, S. L. (2025). Ecosystems: Insights from Complexity. Complexity and Emergence in Market Ecosystems, 1.

Saville, C. E., Smith, H. K., & Bijak, K. (2019). Operational research techniques applied throughout cancer care services: a review. In *Health Systems* (Vol. 8, Issue 1). DOI: 10.1080/20476965.2017.1414741

Schweiger, E., & Tomiak, K. (2022). Researching silence: A methodological inquiry. *Millennium, 50*(3), 623–646. DOI: 10.1177/03058298221083999

Sciberras, M., & Dingli, A. (2023). Research Analysis—Triangulation Approach. In *Investigating AI Readiness in the Maltese Public Administration. Lecture Notes in Networks and Systems* (Vol. 568). Springer., DOI: 10.1007/978-3-031-19900-4_9

Seddon, T. (2023). Comparative education concepts, methods and practices in the emerging anthropocene educational space: From 'measuring the other'to 'supporting the other'? *Comparative Education, 59*(3), 436–457. DOI: 10.1080/03050068.2023.2215643

Shafik, W. (2024a). Artificial Intelligence-Enabled Internet of Medical Things (AIoMT) in Modern Healthcare Practices. In *Clinical Practice and Unmet Challenges in AI-Enhanced Healthcare Systems* (pp. 42–69). IGI Global. DOI: 10.4018/979-8-3693-2703-6.ch003

Shafik, W. (2024d). Data-Driven Future Trends and Innovation in Telemedicine. In *Improving Security, Privacy, and Connectivity Among Telemedicine Platforms* (pp. 93–118). IGI Global. DOI: 10.4018/979-8-3693-2141-6.ch005

Shafik, W. (2024h). IoMT Future Trends and Challenges: Emerging Technologies, Policy Implications, and Research Questions. *Lightweight Digital Trust Architectures in the Internet of Medical Things (IoMT)*, 348–370.

Shafik, W. (2024j). IoT-Enabled Secure and Intelligent Smart Healthcare: Beyond 5G in Enabling Smart Cities. In *Secure and Intelligent IoT-Enabled Smart Cities* (pp. 308–333). IGI Global.

Shafik, W. (2024k). Smart Health Revolution: Exploring Artificial Intelligence of Internet of Medical Things. In *Healthcare Industry Assessment: Analyzing Risks, Security, and Reliability* (pp. 201–229). Springer.

Shafik, W. (2024l). Smart Health Revolution: Exploring Artificial Intelligence of Internet of Medical Things. In *Healthcare Industry Assessment: Analyzing Risks, Security, and Reliability* (pp. 201–229). Springer. DOI: 10.1007/978-3-031-65434-3_9

Shafik, W. (2025a). Artificial Intelligence Assisted Internet of Medical Things (AIoMTs) in Sustainable Healthcare Ecosystem. *Wellness Management Powered by AI Technologies*, 75–101.

Shafik, W. (2025b). Revolutionizing Skin Cancer Diagnosis With Artificial Intelligence: Insights Into Machine Learning Techniques. In *Impact of Digital Solutions for Improved Healthcare Delivery* (pp. 167–194). IGI Global.

Shafik, W. (2025c). Revolutionizing Skin Cancer Diagnosis With Artificial Intelligence: Insights Into Machine Learning Techniques. In *Impact of Digital Solutions for Improved Healthcare Delivery* (pp. 167–194). IGI Global.

Shafik, W. (2024b). Artificial Intelligence-Enabled Internet of Medical Things for Enhanced Healthcare Systems. In *Smart Healthcare Systems* (pp. 119–134). CRC Press., DOI: 10.1201/9781032698519-9

Shafik, W. (2024c). Connected healthcare—the impact of Internet of Things on medical services. In *Artificial Intelligence and Internet of Things based Augmented Trends for Data Driven Systems* (pp. 181–217). CRC Press., DOI: 10.1201/9781003497318-10

Shafik, W. (2024e). Digital healthcare systems in a federated learning perspective. In *Federated Learning for Digital Healthcare Systems* (pp. 1–35). Elsevier. DOI: 10.1016/B978-0-443-13897-3.00001-1

Shafik, W. (2024g). Healthcare and Its Applications in a Smart City. In *Healthcare-Driven Intelligent Computing Paradigms to Secure Futuristic Smart Cities* (pp. 1–35). Chapman and Hall/CRC. DOI: 10.1201/9781032631738-1

Shafik, W. (2024i). IoT-Enabled Secure and Intelligent Smart Healthcare. In *Secure and Intelligent IoT-Enabled Smart Cities* (pp. 308–333). IGI Global., DOI: 10.4018/979-8-3693-2373-1.ch015

Shafik, W. (2024m). The Future of Healthcare: AIoMT—Redefining Healthcare with Advanced Artificial Intelligence and Machine Learning Techniques. In *Artificial Intelligence and Machine Learning in Drug Design and Development* (pp. 605–634). Wiley., DOI: 10.1002/9781394234196.ch19

Shafik, W. (2024n). The Future of Healthcare: Data-Driven Trends and Innovation. In *Advancing Healthcare through Data-driven Innovations* (pp. 166–185). CRC Press. DOI: 10.1201/9781003475392-11

Shafik, W. (2024o). Toward a More Ethical Future of Artificial Intelligence and Data Science. In *The Ethical Frontier of AI and Data Analysis* (pp. 362–388). IGI Global. DOI: 10.4018/979-8-3693-2964-1.ch022

Shafik, W. (2025d). Smart biomedical devices for smart healthcare. In *Machine Learning Models and Architectures for Biomedical Signal Processing* (pp. 421–448). Elsevier. DOI: 10.1016/B978-0-443-22158-3.00017-X

Shafk, W. (2023). Wearable Medical Electronics in Artifcial Intelligence of Medical Things. In *Handbook of Security and Privacy of AI-Enabled Healthcare Systems and Internet of Medical Things*. https://doi.org/DOI: 10.1201/9781003370321-2

Shmatko, O., Borova, T., Yevseiev, S., & Milov, O. (2021). Tokenization of educational assets. *ScienceRise*, *1*(45), 4–10. DOI: 10.15587/2519-4984.2021.232321

Shutova, A. A., & Begishev, I. (2023). *Ethical code for medical robotics activity*. DOI: 10.21202/978-5-8399-0803-1_2023_2_15

Silva, R., Inácio, H., & Marques, R. P. (2022). Blockchain implications for auditing. *The International Journal of Digital Accounting Research*, *22*, 1–25. DOI: 10.4192/1577-8517-v22_6

Singh, A., & Parida, R. (2022). Decision-Making Models for Healthcare Supply Chain Disruptions: Review and Insights for Post-pandemic Era. *International Journal of Global Business and Competitiveness*, *17*(2), 130–141. Advance online publication. DOI: 10.1007/s42943-021-00045-5

Sordan, J. E., Marinho, C. A., Oprime, P. C., Pimenta, M. L., & Andersson, R. (2023). Characterization of Lean Six Sigma projects in healthcare settings: Empirical research. *Benchmarking*, *30*(10), 4058–4075. Advance online publication. DOI: 10.1108/BIJ-03-2022-0183

Sparer, L. (2024). *Token distribution in reward schemes*. BCCA, 91–98. DOI: 10.1109/BCCA62388.2024.10844480

Stake, R. E. (1995). *The art of case study research*. Sage Publications, Inc.

Suan, D. L.Darwin L. Suan. (2024). Token approach in learning motivation. *EPRA IJ Environmental Economics*, *11*(4), 45–56. DOI: 10.36713/epra17411

Sunyaev, A., Kannengießer, N., Beck, R., Treiblmaier, H., Lacity, M., Kranz, J., Fridgen, G., Spankowski, U., & Luckow, A. (2021). Token economy. *Business & Information Systems Engineering*, *63*(4), 457–478. DOI: 10.1007/s12599-021-00684-1

Suyemoto, K. L., Hochman, A. L., Donovan, R. A., & Roemer, L. (2021). Becoming and fostering allies and accomplices through authentic relationships: Choosing justice over comfort. *Research in Human Development*, *18*(1-2), 1–28. DOI: 10.1080/15427609.2020.1825905

Syahrir, I., Suparno, , & Vanany, I. (2015). Healthcare and Disaster Supply Chain: Literature Review and Future Research. *Procedia Manufacturing*, *4*, 2–9. Advance online publication. DOI: 10.1016/j.promfg.2015.11.007

Tan, L. (2018). Meaningful gamification & motivation. *Online Learning : the Official Journal of the Online Learning Consortium*, *22*(2). Advance online publication. DOI: 10.24059/olj.v22i2.1167

Taqafi, I., Belaissaoui, M., Elmrini, Y., & Lakhdissi, M. (2019). Challenges and difficulties related to implement Integrated IS project in Morocco: ERP Projects in Public University case study. *International Journal of Computer Applications*, *178*(51), 33–39. DOI: 10.5120/ijca2019919431

Tartan, C. (2023). Blockchain welfare distribution model. *REGION*, *10*(1). Advance online publication. DOI: 10.18335/region.v10i1.434

Tauveron, C. (1999). *Understanding and interpreting literature in school*. Hachette Éducation.

Tenney, I., Das, D., & Pavlick, E. (2019). BERT Rediscovers the Classical NLP Pipeline (No. arXiv:1905.05950). arXiv. /arXiv.1905.05950DOI: 10.18653/v1/P19-1452

Testimony of Youssef, 13-year-old student. (2025, March 10). Hassan I Middle School, Morocco. [Unpublishedprimary source]

Thomas, G., & Loxley, A. (2022). Groundhog day for inclusive education. *Support for Learning*, *37*(2), 225–243. DOI: 10.1111/1467-9604.12406

Timotheou, S., Miliou, O., Dimitriadis, Y., Sobrino, S. V., Giannoutsou, N., Cachia, R., Monés, A. M., & Ioannou, A. (2023). Impacts of digital technologies on education and factors influencing schools' digital capacity and transformation: A literature review. *Education and Information Technologies*, *28*(6), 6695–6726. DOI: 10.1007/s10639-022-11431-8 PMID: 36465416

Tippong, D., Petrovic, S., & Akbari, V. (2022). A review of applications of operational research in healthcare coordination in disaster management. In *European Journal of Operational Research* (Vol. 301, Issue 1). DOI: 10.1016/j.ejor.2021.10.048

Triputra, J., Nurhayati, N., Choirunisa, R., Pamungkasari, T., & Paramata, V. (2023). Peningkatan Efektivitas Manajemen Kesehatan Melalui Implementasi Teknologi Informasi. *Jurnal Ilmiah Kedokteran Dan Kesehatan*, *3*(1), 220–225. Advance online publication. DOI: 10.55606/klinik.v3i1.2319

Tunisian Ministry of Education. (2021). Theater and citizenship project: Evaluation report 2018–2021. *Pedagogical Bulletin*, *12*(4), 12–34.

UNESCO. Research Report on Digital Transformation of Higher Education Teaching and Learning, ICHEI-April2022. https://en.ichei.org/Uploads/Download/2022-05-16/62820a2a9bceb.pdf

Université Laval. (2016). *Guide des bonnes pratiques de l'enseignement en ligne: Partage d'idées, conseils et suggestions*. https://www.enseigner.ulaval.ca/guide-web/guide-des-bonnes-pratiques-de-l-enseignement-en-ligne

Ur Rasool, R., Ahmad, H. F., Rafique, W., Qayyum, A., Qadir, J., & Anwar, Z. (2023). Quantum Computing for Healthcare: A Review. In *Future Internet* (Vol. 15, Issue 3). DOI: 10.3390/fi15030094

Vitai, Z., & Benke, M. (2021). Learning motivation in CEE students. *Marketing & Menedzsment*, *55*(2), 45–59. DOI: 10.15170/MM.2021.55.02.03

Vygotsky, L. S. (1978). *Mind in society: The development of higher psychological processes*. Harvard UniversityPress., DOI: 10.2307/j.ctvjf9vz4

Walsh, M., Kittler, M. G., & Mahal, D. (2018). Towards a new paradigm of healthcare: Addressing challenges to professional identities through Community Operational Research. *European Journal of Operational Research, 268*(3), 1125–1133. Advance online publication. DOI: 10.1016/j.ejor.2017.05.052

Weller, M. (2020). Open and Free Access to Education for All. In Burgos, D. (Ed.), *Radical Solutions and Open Science. Lecture Notes in Educational Technology*. Springer., DOI: 10.1007/978-981-15-4276-3_1

Wiig, S., Aase, K., Billett, S., Canfield, C., Røise, O., Njå, O., Guise, V., Haraldseid-Driftland, C., Ree, E., Anderson, J. E., Macrae, C., Bourrier, M., Berg, S. H., Bergerød, I. J., Schibevaag, L., Øyri, S. F., Sjøseth, S., O'Hara, J., Kattouw, C. E., . . . Jeppesen, E. (2020). Defining the boundaries and operational concepts of resilience in the resilience in healthcare research program. In *BMC Health Services Research* (Vol. 20, Issue 1). DOI: 10.1186/s12913-020-05224-3

Wijnands, W., & Stolze, A. (2019). Transforming Education with eduScrum. In Parsons, D., & MacCallum, K. (Eds.), *Agile and Lean Concepts for Teaching and Learning : Bringing Methodologies from Industry to the Classroom* (pp. 95–114). Springer., DOI: 10.1007/978-981-13-2751-3_5

Yassine, M., Taouaf, I., Attou, O. E., & Arouch, M. (2024). Structuring Innovation Activities in Moroccan Public Universities: Proposal of a procedures manual. *Salud, Ciencia y Tecnología, 5*, 1066. DOI: 10.56294/saludcyt20251066

Yin, R. K. (2018). *Case Study Research and Applications: Design and Methods* (6th ed.). Sage.

Yu, W., Zhao, G., Liu, Q., & Song, Y. (2021). Role of big data analytics capability in developing integrated hospital supply chains and operational flexibility: An organizational information processing theory perspective. *Technological Forecasting and Social Change, 163*, 120417. Advance online publication. DOI: 10.1016/j.techfore.2020.120417

Zhang, T., Kishore, V., Wu, F., Weinberger, K. Q., & Artzi, Y. (2019). Bertscore : Evaluating text generation with bert. arXiv preprint arXiv:1904.09675.

Ziady, F., Boumahdi, A., & Bezzaz, N. (2024). Scientific and technological research in Morocco: Literature review on constraints and challenges. *International Journal of Accounting, Finance, Auditing, Management and Economics, 5*(5), 309–323. Retrieved from https://www.ijafame.org/index.php/ijafame/article/view/1483

Zweig, J. S., & Stafford, E. T. (2012). Training for online teachers to support student success: Themes from a survey administered to teachers in four online learning programs. *Journal of Online Learning Research*, *2*(4), 399–418.

About the Contributors

Driss Elomari-is a researcher at the University Sidi Mohamed Ben Abdellah Fez, doctor in pedagogical engineering. Currently, he is a lecturer at the National School of Business and Management in Fez. His research focuses on knowledge engineering, pedagogical engineering, qualitative research, economics and knowledge mediation.

***

Nadia Chafiq is a Professor-Researcher, Mathematics, Artificial Intelligence, and Digital Learning Laboratory (MIND-LAB), Department of Computer Science and Educational Science, Hassan II University of Casablanca, Casablanca, Morocco; ORCID 0000-0002-5088-2369, Scopus Author ID 57202151756. E-mail: nadia _chafiq@yahoo.fr.

Rokaya El Gounidi is a Doctoral Candidate, Mathematics, Artificial Intelligence, and Digital Learn-ing Laboratory (MIND-LAB), Department of Computer Science and Educational Science, Hassan II University of Casablanca, STEAM Facilitator and Primary School Teacher, Ministry of National Education, Casablanca, Morocco; ORCID 0009-0008-4289-8759, Scopus Author ID 59918808200. E-mail: rokayaelgouni-di@gmail.com.

Mustafa Kayyali is an ardent advocate for excellence in higher education, driven by a relentless pursuit of quality, recognition, and innovation. Their journey in academia started with a Master's in Quality Management and Evaluation in Higher Education from Universitat Oberta de Catalunya, followed by a Ph.D. in Quality Management from Azteca University. These academic pursuits have laid a strong foundation for their expertise in Accreditation, Quality Assurance, and Higher Education Rankings. As an entrepreneur, researcher, translator, and publisher, they are deeply involved in various facets of the academic world. Their diverse interests encompass Management, Translation, Interpretation, and Academic consulting, contributing to a well-rounded understanding of the industry. Throughout their career, they have made significant contributions to academic literature, with more than 40 published papers and 30 book chapters to my name. Additionally, they take pride in having translated 6 books, bridging language gaps, and promoting knowledge exchange on a global scale. They are committed to fostering positive change in higher education and contributing to its continuous improvement.

Mustapha Khiati is a Professor at the Faculty of Sciences Ben M'Sick, and a member of the Applied Economics and Finance Laboratory (ECAF) at the FSJES of Casablanca.

Wasswa Shafik (Member, IEEE) received a Bachelor of Science degree in information technology from Ndejje University, Uganda, a Master of Engineering degree in Information Technology (Computer and Communication Networks) from Yazd University, Iran, and a PhD in Computer Science with the School of Digital Science, Universiti Brunei Darussalam, Brunei Darussalam. He is also the Founder and Research Director of the Dig Connectivity Research Laboratory (DCRLab) after serving as a Research Associate at the Network Interconnectivity Research Laboratory at Yazd University. Prior to this, he worked as a Community Data Analyst at Population Services International (PSI-Uganda), a Community Data Officer at the Programme for Accessible Health Communication (PACE-Uganda), a Research Assistant at the Socio-Economic Data Centre (SEDC-Uganda), Uganda, an Assistant Data Officer at TechnoServe, Kampala, Uganda, and Asmaah Charity Organization as Cheif Executive Officer. He has hundreds of publications with renowned publishers. His research interests include Artificial Intelligence, Neural Networks, Computer Vision, Smart Agriculture, Digital Health, and Green Computing.

Index